D0478451

MYSTERY OF THE ANCIENT SEAFARERS

MYSTERY OF THE ANCIENT SEAFARERS

DR. ROBERT D. BALLARD
WITH TONI EUGENE

WASHINGTON, D.C.

OCEANIC

A conservator carefully brushes sediment from artifacts found in a wreck dating to the Golden Age of Greece found off the coast of Turkey (opposite).
Preceding Pages: Beside the ruins of a temple to Poseidon, the replica of a Bronze Age Greek galley rounds Cape Sourion, in the Aegean.
Half Title Page: In an ancient mosaic, Romans prepare to cast their nets for fish and eels, delicacies prized by the upper class.

INTRODUCTION

BY SPENCER WELLS, PH.D.

THE SUBJECT OF THIS BOOK IS SO EVOCATIVE THAT, IN many ways, it sums up the spirit of exploration that the National Geographic Society stands for. The thrill of the open sea, the spread of trading networks and cultural interactions, the excitement of reconstructing the distant past with modern scientific methods—what fantastic subjects! The civilizations described here were the world's first internationalists, and they used the sea as a kind of superhighway to expand the limits of the known world. Their legacy lies at the core of what we call Western civilization. Pretty heady stuff. But I'm also excited because this book, and my small role in it, marks a novel combination of two very different fields of scientific research.

Let me explain: I am a geneticist, a job that isn't usually associated with adventure or exploration. Late nights in the lab, yes, but—apart from the occasional broken test tube—not one that creates a frisson of excitement in people's minds. Through a remarkable series of insights over the past 20 years, though, genetics is now able to participate fully in the search for our lost ancestors, their empires, and the rise and fall of civilizations.

There is a parallel between the work of underwater scientists such as Bob Ballard and those of us who tease apart the historical information in the blood we carry in our bodies. On the surface, one ocean, or one drop of blood, looks pretty much like another. But look underneath the surface, use a bit of technology and chutzpah, and amazing things can be revealed. Until someone invents a time machine, the application of these technologies is the best way we have to peer back into the past. Archaeology and genetics provide a portal, a way to relive the lives of our ancestors and witness what their world was like.

As a "gene jockey," my contribution is the ability to trace actual biological relationships. From this—given enough information about history, archaeology, language, and climate—we can trace the spread of our ancestors around the world. This trace can also tell us about the interaction of people in a particular region. For instance, Egyptian historical records describe invasions of the Levant by the Sea Peoples in the 12th century B.C. The archaeological record, on the other hand, suggests that if these invasions did occur, they were probably not hugely disruptive. Can we solve this conundrum with DNA? In particular, can we detect genetic lineages that may have arrived in the Levant from somewhere else, at around that time? If so, where did they come from? And finally, were the Sea Peoples the ancestors of the Philistines or the Phoenicians—and of today's Lebanese doctors, bankers, and market traders?

I've spent the past year working in and around the Mediterranean, trying to tease apart the answers to these and other questions. Retracing the routes followed by the Phoenicians, I've traveled from Lebanon to Tunisia and Spain, onward to Malta and back again to the Levant, looking for the DNA thread that ties these people together. The Phoenicians, once the Mediterranean world's most far-ranging seafarers, have vanished beneath the waves—but I hope their genetic lineages live on. As this book goes to press, our work is still in the early stages, but one thing does jump out of the data—how closely intertwined we all are at the genetic level. It's amazing to think that you or I could be carrying a direct genetic link back to a Sidonian king who lived about 2,500 years ago, or a Tyrian sailor, or a Carthaginian princess—someone in the distant past who binds us together in the present.

This last point gets to the crux of what it is that geneticists do. Our work relies on the cooperation of people alive today—we all carry, hidden in the DNA code of A, C, G, and T, the greatest history book ever written. It is the story of how we came to be. And crucially, it is not simply a fossil, preserving a palimpsest of the past. I am often asked about the relevance of my work to today's world—why should we care about our ancestry? After all, the future seems to be bearing down upon us at an ever-accelerating pace. Surely we have better things to do than spend our time worrying about people who lived hundreds or thousands of years ago? My answer is that everything we are—both biologically and culturally—has been inherited from these ancestors. We are not blank slates, brought into the world devoid of any historical influences. On the contrary, history has enormous relevance for us today. To understand ourselves, our place in the modern world, and—perhaps—where we might be headed in the future, we have to start by looking at the past.

What follows then is a lavish, beautifully illustrated introduction to some of the recent work on these ancient civilizations by one of the world's great explorers, Robert D. Ballard. Bob has been working on ancient shipwrecks in and around the Mediterranean for many years and has made some truly seminal contributions to our understanding of their history. Using some pretty impressive technology, he has plumbed the depths of the Black Sea in search of the great flood, caught tantalizing glimpses of Phoenician shipwrecks, and reevaluated the collapse of the Minoan civilization. These have particular relevance for anyone reading this book since, as someone familiar with written English, we are the inheritors of a long tradition of Mediterranean inventions—not the least of which is the alphabet that the book is written in, developed by the Phoenicians in the second millennium B.C. and later modified by the Greeks and Romans. Read on, and find out what else those long-dead people of the waves did for us—I guarantee that you'll discover something surprising. ■

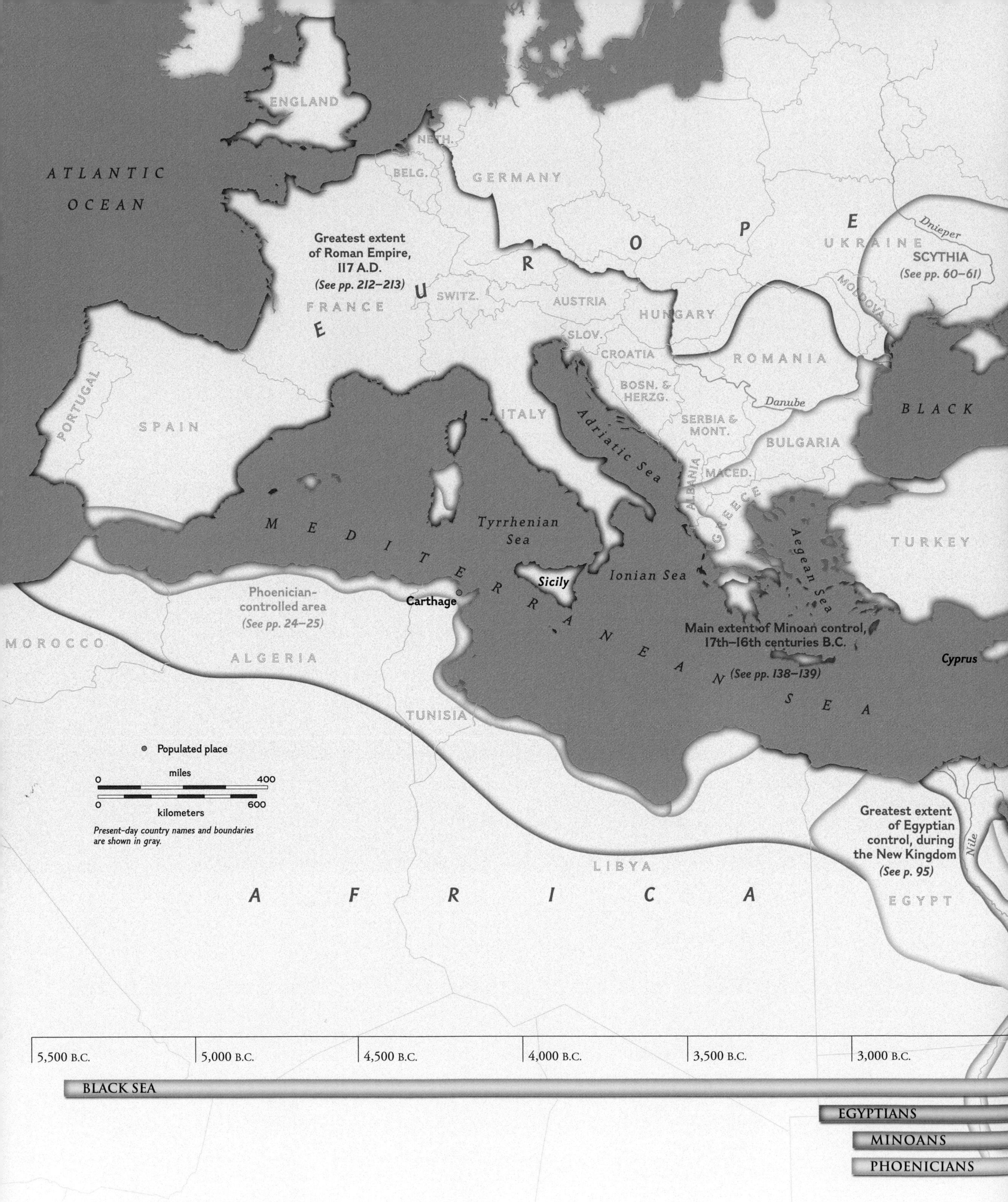

ATLANTIC OCEAN
ENGLAND
NETH.
BELG.
GERMANY
FRANCE
SWITZ.
AUSTRIA
HUNGARY
SLOV.
CROATIA
BOSN. & HERZG.
SERBIA & MONT.
ROMANIA
BULGARIA
UKRAINE
MOLDOVA
PORTUGAL
SPAIN
ITALY
ALBANIA
MACED.
GREECE
TURKEY
MOROCCO
ALGERIA
TUNISIA
LIBYA
EGYPT
E U R O P E
A F R I C A
BLACK
Dnieper
Danube
Nile
SCYTHIA
(See pp. 60–61)
Greatest extent of Roman Empire, 117 A.D.
(See pp. 212–213)
Adriatic Sea
Tyrrhenian Sea
Ionian Sea
Aegean Sea
MEDITERRANEAN SEA
Sicily
Carthage
Cyprus
Phoenician-controlled area
(See pp. 24–25)
Main extent of Minoan control, 17th–16th centuries B.C.
(See pp. 138–139)
Greatest extent of Egyptian control, during the New Kingdom
(See p. 95)
Populated place
miles
0
400
kilometers
0
600
Present-day country names and boundaries are shown in gray.
5,500 B.C.
5,000 B.C.
4,500 B.C.
4,000 B.C.
3,500 B.C.
3,000 B.C.
BLACK SEA
EGYPTIANS
MINOANS
PHOENICIANS

EARLY MARITIME CIVILIZATIONS

3,000 B.C.–500 A.D.

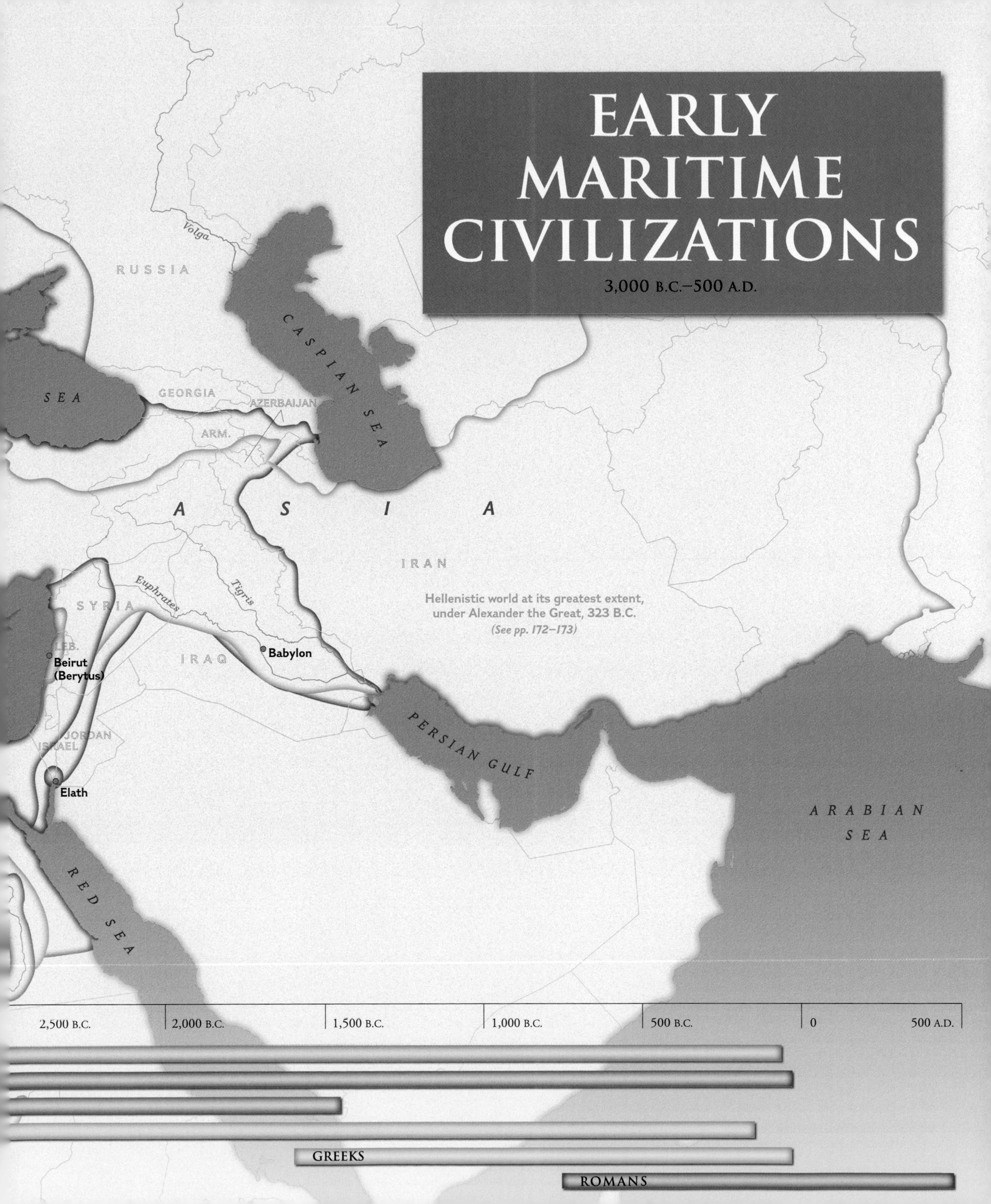

Volga
RUSSIA
CASPIAN SEA
SEA
GEORGIA
AZERBAIJAN
ARM.
A S I A
IRAN
SYRIA
Euphrates
Tigris
LEB.
Beirut (Berytus)
IRAQ
Babylon
JORDAN
ISRAEL
Elath
PERSIAN GULF
ARABIAN SEA
RED SEA

Hellenistic world at its greatest extent, under Alexander the Great, 323 B.C.
(See pp. 172–173)

2,500 B.C. | 2,000 B.C. | 1,500 B.C. | 1,000 B.C. | 500 B.C. | 0 | 500 A.D.

GREEKS
ROMANS

CHAPTER ONE

THE PHOENICIANS

3000 B.C. — 146 B.C.

Horse-headed Phoenician ships carrying logs adorn a wall of the palace of emperor Sargon at the city of Khorsabad; the bas relief dates from the eighth century B.C.

THE PURPLE MEN

BY ROBERT D. BALLARD

OF ALL THE SAILORS IN THE ANCIENT WORLD, THE Phoenicians are to me the most mysterious. As early as 3000 B.C. they were venturing west from their homeland in present-day Lebanon and Syria to dominate the Mediterranean Sea. The Greeks, who coveted the purple cloaks these traders wore, called them Phoenicians—the "purple men"; the Phoenicians referred to themselves as *Kena'ani*—Canaanites.

They were the most daring explorers and mariners in the ancient Near East, yet archaeologists know little of their actual ships and trade routes.

Harvard University archaeologist Dr. Lawrence Stager and I led an expedition in the summer of 1999 that will add a new chapter to the history and archaeology of Phoenician trade. Two years before, a nuclear research submarine had located three targets off the coast of southern Israel. We had returned with sensitive sonar to survey and discover the actual sites and the remotely operated vehicle (ROV) *Jason* to explore them and extract artifacts that would help date and identify the wrecks. Expectation built in our surface support ship, the *Northern Horizon,* as Jason's cameras relayed flickering images of the first wreck to our onboard video monitor. Tension grew as a picture of an old-fashioned steel anchor and rusty chain resolved on the screen. There was no magic moment here; this was an 18th- or 19th-century sailing vessel, not the ancient Phoenician "round boat" we had hoped to see.

With mounting disappointment, we sent Jason to the second target, about a mile and half away. Every one of the 49 scientists and graduate students on the mission waited anxiously. Moments later, to whoops of triumph, an image of hundreds of stacked amphorae appeared on the sonar screen. The hull above the mud line had vanished, eaten by wood-borers, but the dark heaps of amphorae marked the cargo holds, and a looser collection of artifacts delineated the bow and the stern of a vessel about 48 feet long. Stager validated our hopes promptly: The amphorae were Phoenician.We were looking at the first Iron Age ship ever discovered in the deep sea, not near the shore—a vessel that sailed some 800 years before the birth of Christ.

Two days later, Jason's sonar recorded another hit—the shimmering image of a second and very similar merchantman. Like the first, it had settled unbroken and upright on the soft seafloor and finally succumbed to wood-borers. It, too, carried a cargo of wine—at least 400 amphorae—and at 59 feet long proved to be the largest preclassical shipwreck ever discovered.

Maritime research, sadly, is as subject to budget and scheduling constraints as any Fortune 500 company. Our time and money were running out. We would return to the ships we had named the *Tanit* and the *Elissa* after Phoenician figures. In our short time, however, we had discovered much. We had established that the Phoenicians were daring mariners who sacrificed the reassuring presence of land for speed; they did not, as some scholars have believed, hug the shoreline and try to beach their ships at night. Instead, they braved the deep sea to sail the shortest course to a distant objective.

These two ships probably sailed south from the Phoenician port of Tyre, stopped at the ancient city of Ashkelon, and left there to deliver their cargoes of wine en route to Egypt or the north African city of Carthage, some 1,600 miles away. They may have been part of a Phoenician trading fleet when an unexpected east wind from the desert swept them off course, where they flooded and foundered.

The two wrecks offer much more to marine archaeologists. Perhaps some wood of the hulls remains buried under the soft sediments of their burial grounds. We had planned to conduct further research during the summer of 2003, but the Egyptian government rescinded our permit due to military concerns. Politics as well as money hampers marine archaeology. With increasingly advanced technology that allows us access to deepwater sites, however, we will find many of the treasures that the ocean preserves. We will return to the Phoenician wrecks and are already planning a future expedition to locate Phoenician ships lost in international waters. ■

"

THE PHOENICIAN SHIPS
HAD THE BEST
SAILORS IN THE FLEET,
AND THE SIDONIANS THE
BEST AMONG
THE PHOENICIANS.

"

HERODOTUS, BOOK 7, CHAPTER 96

FINDS AWAIT

In 1999 Robert Ballard, Larry Stager, and their team explored two Phoenician ships off the coast of Israel. On the support vessel (above), Ballard manned a control stick as the team studied videos relayed by the remotely operated vehicle (ROV) Jason 2,700 feet below. Jason recorded more than 800 images, which were assembled into a composite of the amphorae from the wreck (right); another layer of jars may lie buried beneath. Ballard plans to return to the site for further exploration.

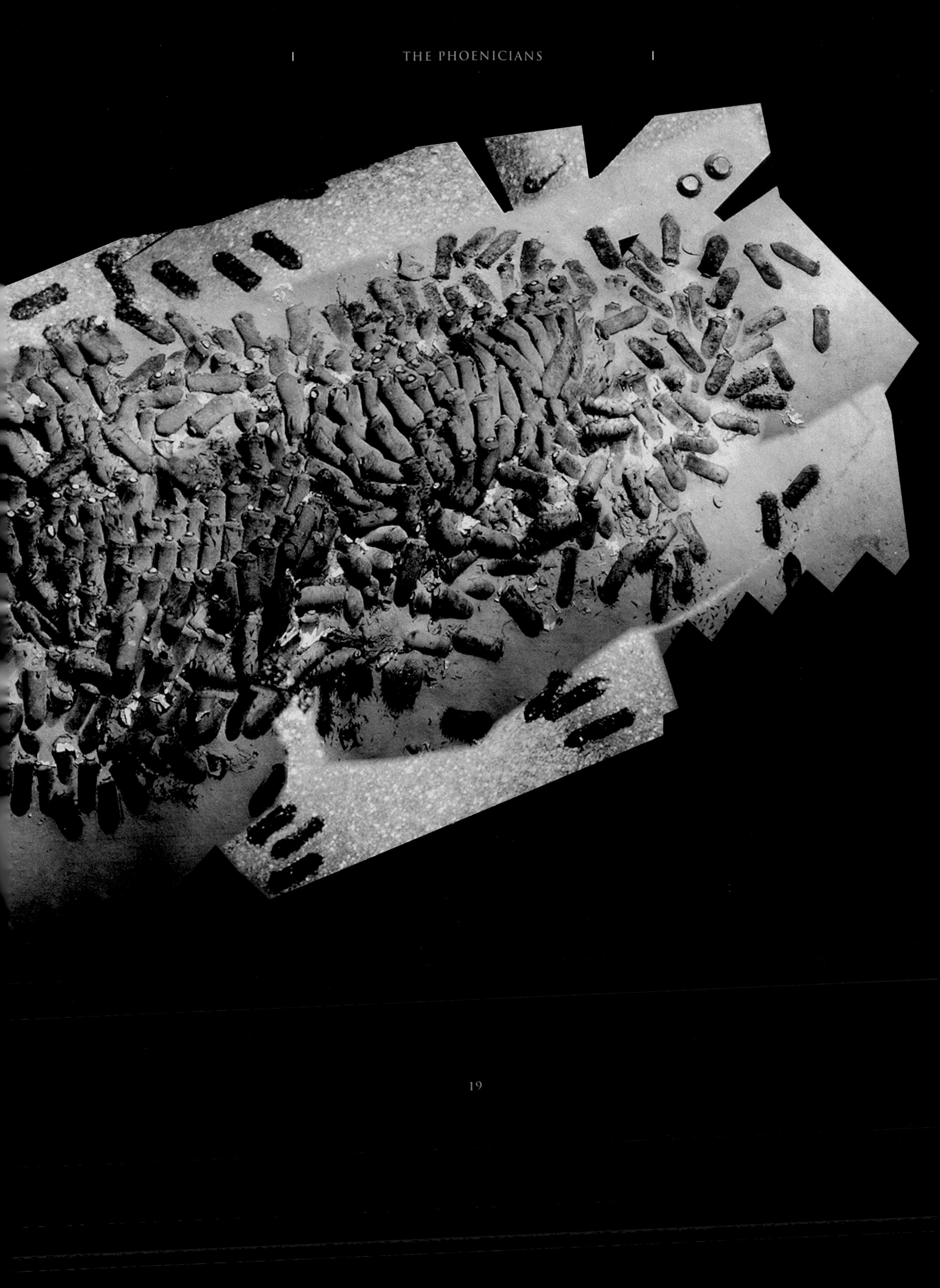

BEATEN GOLD:
A funerary mask from the fifth century B.C. unearthed near present-day Tortosa, Syria, attests to the craftsmanship of Phoenician goldsmiths.

MASTER MARINERS OF THE ANCIENT WORLD

THREE THOUSAND YEARS AGO, AS AGRICULTURAL villagers built burial mounds in the Ohio River Valley and the Olmecs developed Mesoamerica's first true civilization in the steamy lowlands of Central America's Gulf Coast, Canaanite traders from the area now called Lebanon and Syria commanded the waters of the Mediterranean Sea. The Greeks named them Phoenicians, from *phoinix*, or purple, because these traders were known for their purple cloaks.

Over time these so-called Phoenicians established a trading network that stretched from Spain to Persia. Beginning about 1200 B.C., Phoenician ships ruled commerce in the Mediterranean, a supremacy that lasted for more than a thousand years. They traded their native cedars for Egyptian papyrus, supplied the Israelite King Solomon with lumber, and lent their engineering expertise to building his temple and palaces in Jerusalem. They ventured even beyond the Mediterranean and made the first recorded circumnavigation of the African continent. To keep track of their far-flung business interests, the Phoenicians developed a system of writing that evolved into our Western alphabet.

Yet we know very little about these greatest maritime merchants of the ancient world. Few Phoenician written records survive, and those are mostly votive or funerary inscriptions on stone monuments. The Phoenicians kept records on papyrus, which rotted in the damp coastal climate of their homeland, and on potsherds, of which only few have survived. Most of our knowledge of the Phoenicians comes from the records of rivals and enemies. Crumbling walls are all that remain of many Phoenician trading posts; others, like the one at present-day Beirut, lie many layers beneath the city. In Beirut, older than Rome and the most spectacular of Phoenician cities still extant, the settlement of Berytus lies far below Ottoman, Muslim, Crusader, Byzantine, Roman, and Greek remains.

Only in the last hundred years have discoveries added to our knowledge of the Phoenicians and their maritime civilization. Archaeologists in the Mediterranean have amassed a growing collection of artifacts—from land and sea. At Byblos, oldest and richest of Phoenician cities, archaeologists have been excavating since 1925. Yet no intact buildings, no huge arches or columned temples rise above foundation walls overgrown with weeds and grass. Excavations began in Tyre in 1947, where, about 950 B.C., King Hiram joined two islands and built on them a citadel in the sea. Sarepta, the only Phoenician city in Lebanon not overlain by a modern town, was discovered under a wheat field in 1970. Excavations by Dr. James B. Pritchard of the University of Pennsylvania Museum have uncovered nine different cities there, representing all of Phoenician history. Various trades and industries were concentrated in separate quarters, as is the pattern in many towns in the Near East today. Kilns and masses of potsherds reveal the history of centuries of pottery making. Stone cylinders indicate that Sareptans crushed olives for cooking oil. In a one-room temple priests presented altar offerings to their god Baal.

Although archaeological evidence added knowledge of their maritime past, during a century of research not one Phoenician ship was discovered in a harbor or ancient dockyard, and no drawing of a vessel was found. Finally, in 1958, a local sponge fisherman reported the existence of a wreck in

CARTHAGE, *greatest Phoenician colony and its western Mediterranean capital in North Africa, looms in this imagined view by J.M.W. Turner, 1815, in the National Gallery of London.*

a hundred feet of water off Cape Gelidonya, in southern Turkey. In 1960 American archaeologist George Bass and journalist Peter Throckmorton discovered that marine worms had eaten nearly all of the hull of the ship; it sank on a hard, rocky bottom where little sand or mud covered and preserved it, and only a few bottom planks remained. The ship's cargo lay where it had originally settled; by mapping each object in the wreck, Bass and Throckmorton estimated that it had been about 50 feet long. Ingots of copper and tin—raw material for bronze—comprised most of the cargo. The divers also raised a stone anvil, hammers, and smithing tools. A small collection of personal belongings—a seal for signing clay tablets, carved scarabs, some stone tools, and graduated weights—proved to be of Canaanite origin, as did the ship's anchor found decades later. Bass determined that it was a proto-Phoenician ship owned or leased by a trader-craftsman. Carbon dating of twigs and branches used to pack the cargo, as well as the style of the objects recovered, confirmed that the wreck dated from about 1200 B.C., the oldest shipwreck found up to then.

In 1984 George Bass and Cemal Pulak, with a team from the Institute of Nautical Archaeology at Texas A&M University, began excavating a second—and even older—late Bronze Age wreck. This ship, found off Ulu Burun, a cape west of Gelidonya, could be dated to about 1300 B.C., the oldest wreck found so far. The cargo included copper ingots from Cyprus and tin that may have come from the Taurus Mountains in Turkey. Most of the other items originated in the Levant. Canaanite amphorae, Syrian flasks and bottles, and a bronze Canaanite statuette of a goddess covered with gold leaf indicated that the ship probably sailed from the Near East as well. Stone anchors of a type similar to others found on the coast of the Phoenician homeland provided clues to the ship's home port. Based on tons of material excavated from the wreck, Bass concluded that the ship was proto-Phoenician and was probably sailing from east to west, possibly headed to Greece or even Egypt with a royal cargo of tribute.

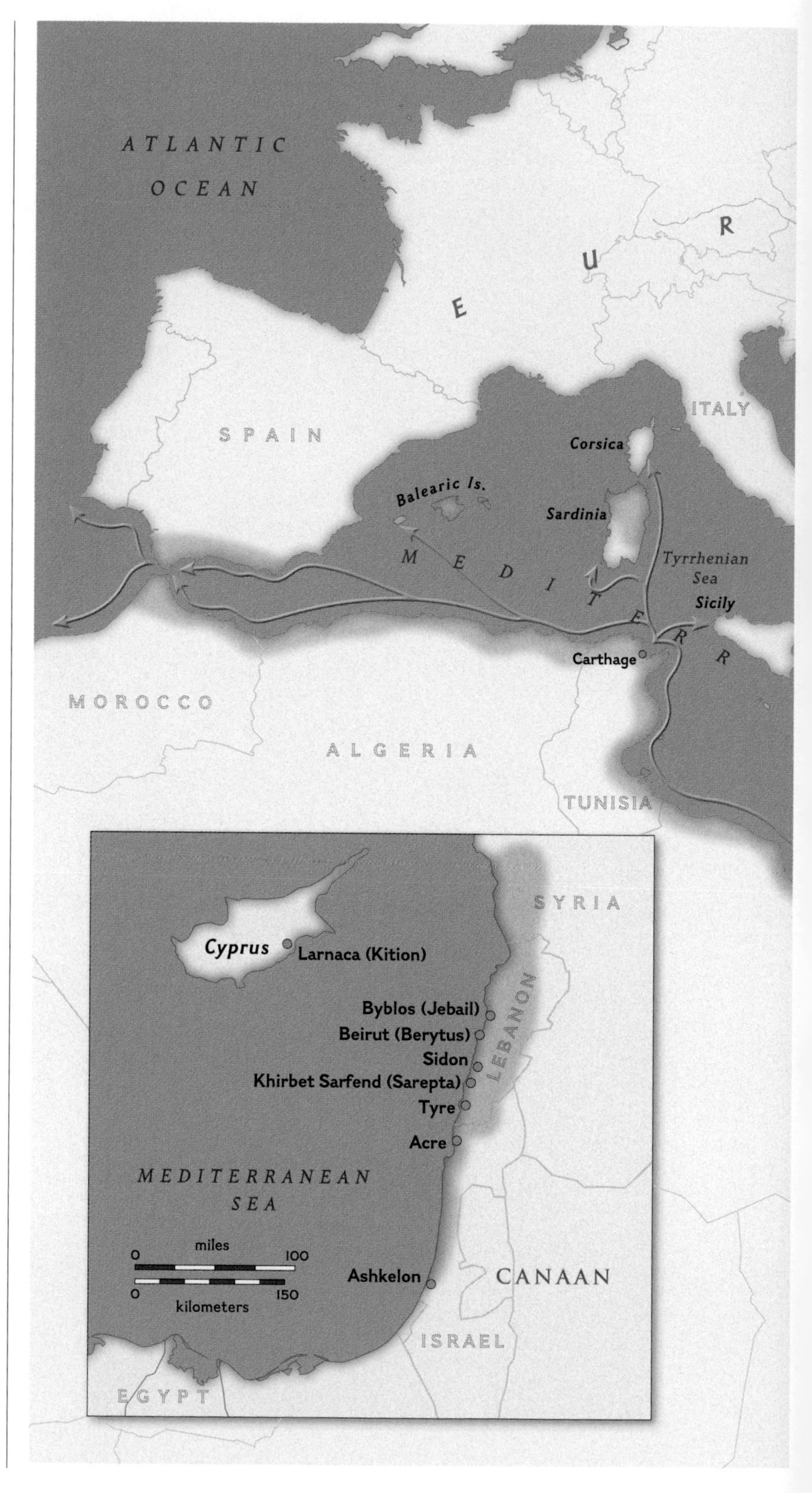

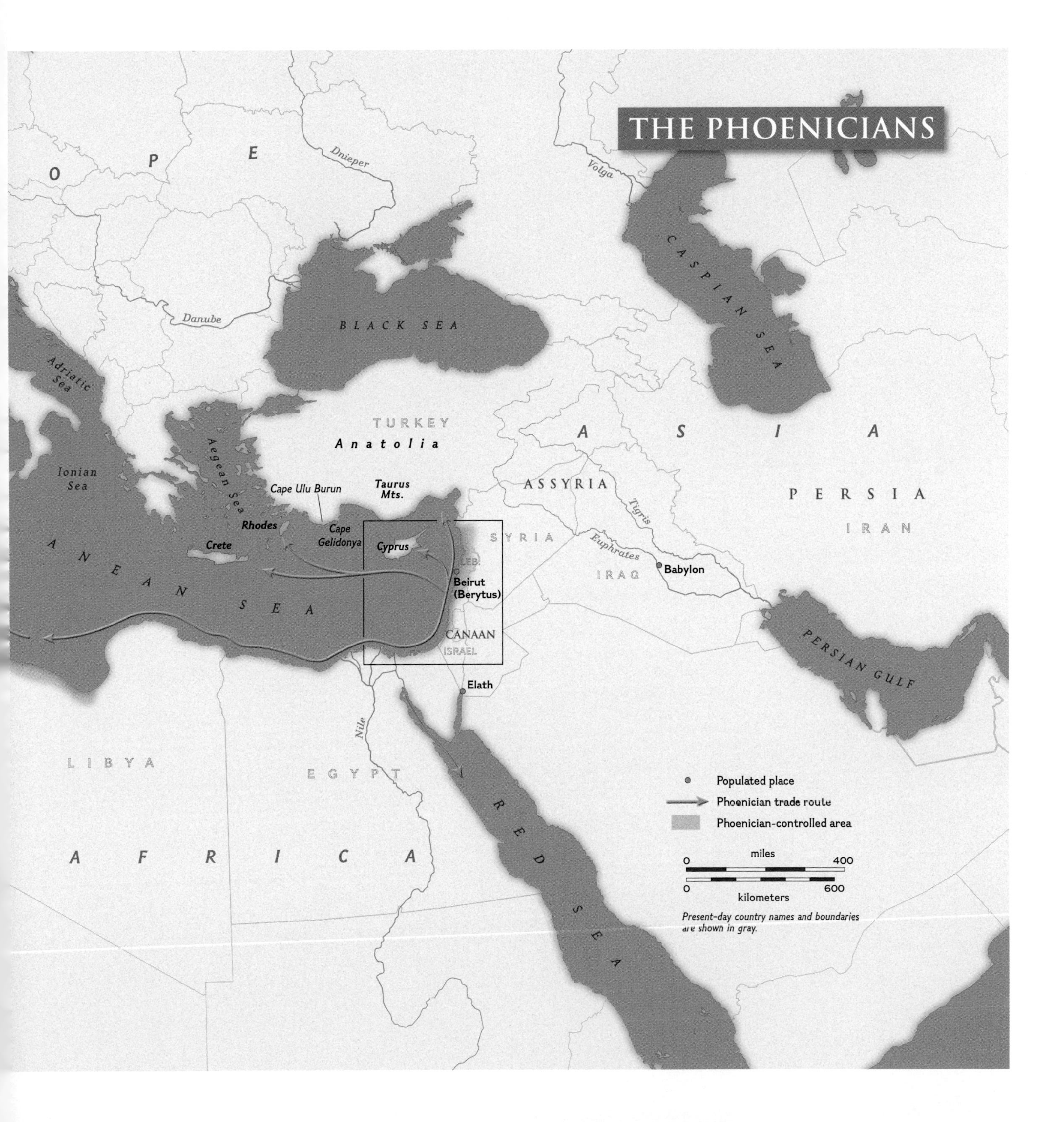
THE PHOENICIANS
O P E
Dnieper
Volga
CASPIAN SEA
Danube
BLACK SEA
Adriatic Sea
TURKEY
Anatolia
A S I A
Ionian Sea
Aegean Sea
Cape Ulu Burun
Taurus Mts.
ASSYRIA
Tigris
PERSIA
IRAN
Rhodes
Crete
Cape Gelidonya
Cyprus
SYRIA
LEB.
Euphrates
IRAQ
Babylon
Beirut (Berytus)
CANAAN
ISRAEL
PERSIAN GULF
Elath
Nile
LIBYA
EGYPT
RED SEA
AFRICA
Populated place
Phoenician trade route
Phoenician-controlled area
miles
0
400
0
600
kilometers
Present-day country names and boundaries are shown in gray.

ASHKELON, *a major seaport since 3500 B.C., now rises above the waves in present-day Israel. Babylon burned the city in 604 B.C. The Phoenician city-state of Tyre restored the 150-acre port, but it was destroyed again by the Mamluks. Until comprehensive excavations began there in 1985, it lay under centuries of rubble.*

SINCE THE IRON AGE THE WORLD HAS KNOWN the coastal Canaanites as Phoenicians. In Neolithic times these people—Semitic tribes of farmers and herders—had wandered into the Levant. By the third millenium they had settled the cities of Byblos, Tyre, Sidon, and Arwad along the coast of the eastern Mediterranean. They called themselves "Canaanites," or more frequently identified themselves by the city or city-state from which they came, i.e. Tyrian, Sidonian, and the like. The area was fertile, rich in rivers, springs, and shallow offshore fishing beds, hemmed in on the east by forested mountains. The Mediterranean offered new opportunities, and soon the Canaanites turned to maritime trade. They established a far-ranging mercantile network and sailed their ships along the eastern seaboard. During the first half of the second millennium B.C., they branched out farther afield to trade with Crete, Cyprus, Egypt, and southern Canaan.

In the first millennium B.C. Greeks gave the name "Phoenicia" to the narrow coastal strip of modern Lebanon and northern Israel, from Tyre in the north to Acre in the south. In the early 12th century B.C. power had shifted in that area. After several hundred years, Egyptian domination of the Canaanites began to flag under Pharaoh Ramses III. A half dozen groups known as Sea Peoples, among them the Philistines, invaded the eastern Mediterranean and threw the area into turmoil.

The Phoenician cities, each with its own hereditary king, were extremely competitive. Phoenicia, never a nation or an empire, consisted of a chain of coastal cities that dominated Mediterranean trade. The cities of Tyre and Sidon gained in importance; together with Byblos and Arwad they each became centers of political and commercial power.

At the crossroads of the ancient world, Phoenician cities lay at the end of inland trade routes leading from Assyria and Persia to the Near East. In addition, vast forest reserves of cedar and fir blanketed mountain slopes east of its cities. Early ships carried logs to Egypt, and returning merchantmen imported grain, linen, gold, and papyrus. Phoenician cities grew as distribution centers. Byblos—present-day Jebail—was undamaged by the onslaught of the Sea Peoples and had a natural harbor. Phoenician trade was strengthened by construction of harbors at Tyre and Sidon. The tiny island of Aradus—now Arwad—became a critical port on the route to the Aegean and southern Anatolia. Like Tyre and other Phoenician cities, Arwad was an island fortress—a town ringed with turreted walls offshore of the mainland or connected to it by a causeway.

Political alliances strengthened the physical defenses of the Phoenician city-states. A treaty between King Hiram of Tyre and Israelite King David in the first millennium B.C. seems to have fortified the city's access to inland trade routes. David won Phoenician neutrality during wars with the Philistines by according Tyre commercial advantages over its rivals. A joint enterprise between the next Israelite king, Solomon, and Hiram resulted in the construction of a port, now Elath, on the Gulf of Aqaba as a base for trade in the Red Sea. As their Assyrian neighbors grew in might, Tyre and Sidon held them at bay by paying tribute.

Most ancient trade routes followed coastlines or hopped between islands; rarely did sailors navigate far offshore. The Phoenicians, it is said, though they usually ventured forward with caution and stayed near shore, were the first mariners to undertake regular trips beyond sight of land and to travel at night, steering by the stars. Their contemporaries did not attempt the long voyages that made the Phoenicians famous.

Their original trade in cedars of Lebanon, shipped from Byblos, was soon augmented with fabric from Tyre and Sidon. Workmen in these cities took advantage of large offshore beds of murex sea snails to produce a rare, purple dye coveted by kings and emperors. Heaps of shells are still visible at the remains of ancient Sidon. Phoenician traders in purple cloaks pioneered sea routes to the western

Mediterranean. Their ships carried embroidered clothing from Mesopotamia, linen from Egypt, wool from the East, copper from Cyprus, and gold from Ophir—a mysterious place, possibly in Africa or Arabia. The search for metal, especially silver and tin, impelled the Phoenicians westward. By the first millennium B.C. they had sailed past Sicily, for centuries the western limit for early navigators. One trade route traced the coasts of that island, Sardinia, and the Balearic Islands. Another followed the coast of northern Africa.

The Phoenicians grew skilled as craftsmen as they excelled in commerce. Artisans from Tyre and Sidon produced glass beads and ships for export. Phoenician craftsmanship is even mentioned in the Bible—a word deriving from the name Byblos—and praised by Greek epic poet Homer, and their wares have been excavated from one end of the Mediterranean to the other. They were masters at making glass, weaving luxurious cloth, dyeing wool in the color purple, fashioning fine wood furniture, and working bronze, gold, and ivory. Although they usually constructed their own houses of stone or brick, Phoenician carpenters and cabinetmakers were so famous that Israelite King Solomon imported them to work on his temple and palaces in Jerusalem.

Colonization followed in the wake of commerce. Traditional sources indicate that the Phoenicians sailed as far west as Spain during the late second millennium B.C. The earliest archaeological evidence of their westward explorations, the Nora Stone excavated in Sardinia, contains a Phoenician inscription that dates to the ninth century B.C. At the same time, Phoenician city-states were establishing colonies on the island of Cyprus.

TREASURES OF ASHKELON: *A bronze bull calf, plated with silver, and the remains of the pottery shrine that held it were unearthed in 1990. The four-inch-high calf represented the divine mount of the gods.*

By 850 B.C. Phoenician settlers had converted an existing Mycenaean site at Kition (modern Larnaca) into a temple to their goddess Ashera, known by her Greek name as Astarte. Remains of votive pits provided evidence of their sacrifices and cult practices. Although Greek merchants also plied the Mediterranean, the Phoenicians were more skilled and daring and sailed much farther west. By 800 B.C. their search for tin, which when alloyed with widely available copper produced bronze, had drawn them beyond the Straits of Gibraltar, known to the ancient world as the Pillars of Hercules. They ventured north into the Atlantic and founded their westernmost outpost, now Cadiz, on the western coast of present-day Spain. More than 2,500 miles from their home ports, the Phoenicians founded colonies on the Atlantic and northern coasts of Africa.

Northeast of modern-day Tunisia, where the Mediterranean tapers to its narrowest point, the Phoenician city-state of Tyre founded Carthage, a city that would someday rival Rome. The traditional date for the founding

SKILLED SAILORS *bring a Phoenician merchant vessel into the harbor of Alexandria, Egypt, one of the busiest ports in the ancient world, in a modern water-color (left). The deep, rounded hulls of Phoenician merchant ships afford-ed plenty of cargo space. Phoenician warships, built for speed, were more streamlined in shape, but both types of vessels relied on oars and a single sail. A golden pendant from about 400 B.C. (above) shows the details of a merchant ship, a symbol of the importance of commerce to Phoenician civilization, and is an example of Phoenician superior craftsmanship.*

FACES *reflect the skill and diversity of their Phoenician craftsmen. Tharros, on the island of Sardinia, yielded a grimacing terra cotta mask (opposite). From Carthage, one of molded and painted terra cotta (above) was used in death rites. A glass pendant of a bearded head (left) from Carthage probably served as jewelry or a charm to ward off evil spirits.*

of Carthage is 814 B.C.; the earliest archaeological evidence dates from 50 years later. By 600 B.C. the Phoenicians had established a small colony south of Gibraltar on the Atlantic coast of present-day Morocco. According to Greek historian Herodotus, they also sailed around the African continent.

Part of the Phoenicians' success rested in their skill and audacity as sailors; part rested in their ships. Long experienced in the timber trade, the Phoenicians became master shipbuilders. Until the 20th century, when scuba equipment enabled divers to explore the depths, archaeologists had no idea how the ancients built their boats. The Phoenicians fastened the planks of a hull together with mortise and tenon joints, a practice then unknown to other shipbuilders of the region. When the hull was completed, they inserted a complete set of frames into the shell to stiffen and strengthen it. Phoenician shipyards, greatest of their time, produced heavy ships, one hundred feet or more in length. The ships were galleys, long, low wooden vessels propelled chiefly by oarsmen. Galleys originally had only one line of rowers on each side. The deep, broad-beamed hulls were rounded at stem and stern. Many were adorned with a horse's head or other ornament on the upraised prow. One type—wide, heavy, roomy, and strong—transported goods and passengers. These trading vessels—referred to in ancient records as "round ships"—resorted to oarsmen when wind on the single square sail was insufficient to make progress. Experts estimate that the merchantmen traveled about 30 miles in a day.

To maintain their home cities against raiders as well as protect their merchantmen and colonies, the Phoenicians developed warships. Light and slender, these streamlined galleys transported dispatches and personnel. A battering ram sheathed in bronze protruded from the prow of war galleys at the waterline. At first, a single line of rowers on each side of the ships provided maneuverability as well as speed. Then biremes, ships with two staggered levels of oarsmen, were developed. By the fifth century B.C. triremes, galleys with rowers on three levels, a mainsail, and a small foresail, had replaced biremes.

The speed and power of these ships helped little against the rising might of Assyria east of the Phoenician homeland. Although Tyre, Sidon, and other city-states continued to pay tribute to the Assyrians, their demands and violence increased during the eighth century B.C. In about 700 B.C. the Assyrian king Sennacherib invaded Tyre—safe for centuries on its nearly unimpregnable island—and burned the city. Wall reliefs from Sennacherib's palace at Nineveh contain the fullest extant representation of the Phoenician navy, revealing the inward-curving stern, high deck, and long ram of their war galleys.

Assyria continued to exact tribute from the Phoenician city-states until Babylonia assumed control. Tyre was assaulted again by Babylonian king Nebuchadnezzar in 586 B.C. After a 13-year siege of the island-city, Nebuchadnezzar's army withdrew, and Tyre escaped destruction. Some 50 years later, the Persians conquered Babylonia, and the victors took over control of Phoenicia.

As Assyrians, then Babylonians, then Persians invaded the Phoenician homeland, their settlement at Carthage, on the coast of northern Africa, became a haven for growing numbers of refugees. Unlike Greek settlements, established to grow and become independent, most Phoenician colonies remained small and were dependent on their founders for centuries. Carthage was an exception: It grew in population and power. Maybe the Phoenician population was too small to produce rapid growth in her outposts, and Carthage grew as a result of unrest in homeland city-states. Or perhaps the city, whose name derives from the Phoenician words *Kart-Hadasht*—"new city"—was destined to become a power in the western Mediterranean. Carthage, like other Phoenician city-states, grew wealthy based on trade. Archaeological remains show that the city traded extensively with the Greeks in Sicily and the Aegean

THE STORY OF PRINCESS ELISSA AND THE FOUNDING OF CARTHAGE

In the eighth century B.C., King Pygmalion of Tyre ruled his city with an iron hand and was angered by the priests who threatened his authority. When Pygmalion murdered the high priest Sychaeus, the husband of his sister, Elissa, the princess fled from Tyre. Taking the king's treasure with her, Elissa escaped with priests and temple maidens of the goddess Astarte to the island of Cyprus.

The group sailed farther west to the Gulf of Tunis, on the coast of northern Africa. A local prince, Iarbas, offered Elissa land on which she could build a new home, but she could have only as much territory as an ox hide would cover, Iarbas allowed. The widow Elissa, bright as well as beautiful, cut an ox hide into narrow strips and tied them end to end. Then she carefully marked off an entire hilltop on a bay for her city. That hill still exists today in modern Tunisia and is called Byrsa, which means "ox hide." On this hill Elissa founded Carthage, the city that would one day rule the western Mediterranean and rival Rome in might and prestige.

Brown tones in a grisaille painting by 15th-century Italian artist Andrea Mantegna lend heroic proportion to the founder of Carthage.

The Phoenician myth of the founding of Carthage was altered at the same time it was immortalized many years later in the first century B.C. The greatest of the Roman poets, a farmer's son, Publius Vergilius Maro—Virgil—used the story of Elissa in his epic, the *Aeneid.* Virgil called the Tyrian princess Dido. The poet spent a decade writing the *Aeneid,* and on his death the Roman Emperor Augustus saved the epic even though the author begged to have it destroyed because it needed many more years of work.

The poem chronicled the wanderings of the Greek hero Aeneas, said to be the son of a god, who would become the founder of the Roman nation. After they had fled the burning city of Troy, Aeneas and

A 15th-century oil by Lieberale da Verone immortalizes Dido's suicide at the loss of Aeneas.

his band of warriors roamed the Mediterranean for seven years in search of a new homeland. Sailing west from Crete, they rounded the island of Sicily when they were caught in a storm unlike any before or since. Juno, wife of Jupiter, supreme ruler of all the gods, initiated the stupendous gale. Waves so great that their crests touched the stars and gulfs between them so deep that the ocean floor lay bare battered Aeneas's ships. Finally Aeolus, King of the Winds, calmed the sea, and the Trojans beached their vessels on the north coast of Africa.

In Carthage, Aeneas met Queen Dido, who fell madly in love with the Trojan hero. For some time, the two were happy. Aeneas strolled the streets of Carthage in a robe of purple embroidered with gold especially made for him by his lover.

Jupiter, however, grew impatient and reminded Aeneas that he must leave Carthage to fulfill his destiny of founding the Roman nation. Prompted by the ruler of heaven himself, Aeneas gathered his men and prepared for departure. Dido, disbelieving, begged him to stay, but to no avail. The queen cursed her Trojan lover and his descendants—all future Romans—swearing that they would always be enemies of Carthage, then killed herself with a sword Aeneas had given her. The Trojans sailed that very night. When Aeneas glanced back, he saw the walls of Carthage lit by the leaping flames of a huge fire—the glow of Dido's funeral pyre.

Sea. By 700 B.C. the African settlement had assumed leadership of western outposts and colonies, and by 514 B.C. Carthage was clearly an independent city-state bound to its founding city of Tyre only by tradition and religion.

Carthaginians, like other Phoenicians, worshiped a triad of deities. Their principal god was known, depending on the location, as Baal or Melqart, son of El and his wife Ashera, the goddess of fertility. The young god Baal, often mentioned by the Greek name Adonis, symbolized the cycle of the seasons in his yearly death and resurrection. Phoenicians feared their gods and made sacrifices to them in great temples and open-air sanctuaries throughout the Mediterranean.

A number of primitive features, notably the practice of child sacrifice, characterized Carthaginian religion. To fulfill the vow made to their gods, priests cut the throats of infants and placed them into blazing fires as an offering. Excavations in the 1920s and later in the 1970s unearthed thousands of urns containing the charred bones and ashes of children. Carthaginians practiced such rites for seven centuries; archaeologists have also found evidence of child sacrifice in Sardinia and Sicily, but as yet have uncovered no such grisly evidence in the Phoenician homeland. (Excavations begun in Tyre in 1997 led to rumors of a Tophet—a shrine where sacrifices were performed—but it turned out to be a necropolis of cremated adults and children of the ninth through seventh centuries B.C., with no evidence of sacrifice.)

Carthage was a rough environment in other ways. The city was notorious for its exploitation of non-Phoenician subjects, a practice that allowed it to raise numerous armies and navies. Conscripting rowers from other settlements as well as it own, Carthage regularly manned fleets of up to 200 ships.

In the fifth century B.C. Carthaginians established their own colonies. At the western tip of Sicily they supported a city, known now as Motya, originally founded in the late eighth to early seventh century by earlier Phoenicians. It was a typical Phoenician site—a small island close to shore with good anchorage. About 580 B.C., when Greek settlements in Sicily tried to destroy Motya and the nearby Phoenician settlement of Panomus (Palermo), Carthage led the defense of the threatened outposts. A Carthaginian naval victory over the Greeks off Corsica in 535 B.C. kept the Greeks from trading with parts of southern Spain where Phoenicia had trading interests.

In 1969 a dredge clearing the shallows off Motya brought up the planks and ribs of an ancient shipwreck. British undersea archaeologist Honor Frost discovered many wrecks embedded in the bottom. The stern post of one rose from the sand to within eight feet of the surface. Carbon dating of the blackened and fragile timber as well as pottery found in the wreck indicated that the ship was built near the middle of the third century B.C. The hull timbers were newly cut and fitted; branches used to cushion ballast stones still had leaves attached. Frost determined that the vessel, new when it sank, was probably a warship. The first Phoenician ship raised from the sea, it carried no cargo and went down with the entire squadron in a storm or in a battle.

AFTER BABYLON FELL IN 539 B.C., PHOENICIAN fleets played a vital role for the Persians as they sought to expand their empire and to humble Greece. From the time of Israel's King Solomon, Phoenician fleets sailed on behalf of various nations that lacked navies. Phoenician warships fought for Persian King Darius and, upon his death, his son Xerxes. At Salamis, near Athens, in 480 B.C., Phoenician squadrons formed the heart of Xerxes's fleet in the greatest sea battle of ancient times. The king of Sidon served as an admiral of the Persian fleet, and ancient historians recorded that the Phoenicians fought with skill and bravery. The commanders of the Greek triremes, though outnumbered three to one by similar Persian vessels, used cunning and geography to rout their enemy in a narrow passage, a defeat that not only humbled the Persians but

OLDER THAN ROME: *Modern-day Beirut, Lebanon, rests atop the ruins of settlements from several civilizations. Remains of the original Phoenician settlement of Berytus underlie the crowded capital, seen here in an aerial view.*

also foreshadowed the end of Phoenician sea power.

During the height of Persian influence in the eastern Mediterranean, the Phoenician cities still prospered for a while, as established by a team of archaeologists, headed by Lawrence Stager of Harvard University. Since 1985 the archaeologists have been excavating an ancient site called Ashkelon, a major seaport dating as far back as 3500 B.C. on the coast of Israel. A strategic site on the trade routes from Turkey and Syria to Egypt, Ashkelon witnessed the rise and fall of civilizations from the time of the Canaanites through Philistine, Greek, Roman, Byzantine, and Crusader invasions. Stager has unearthed layers representing all these different cultures.

BRONZE-AGE *walls remain of Byblos (opposite), oldest and richest of Phoenician homeland cities; beyond them rise the ruins of a later Roman colonnade. The tomb of a king contained an eight-inch-long gold pectoral ornament (above), which is now at the Louvre.*

In 604 B.C. the Babylonian King Nebuchadnezzar burned Ashkelon to the ground, apparently to discourage Phoenician cities from making alliances with rival Egypt. Stager has unearthed a layer of charred wood and debris that testifies to the city's total destruction. When the Persians conquered the Babylonians some 75 years later, the victors urged their Phoenician allies to rebuild Ashkelon; Tyre restored it to its former glory, but the city was destroyed again by the Mamluks in A.D. 1270. Until Stager's arrival, the 150-acre city lay buried under centuries of accumulated rubble.

Stager's excavations indicate that Ashkelon, like other port cities, hosted a diverse population—possibly 15,000 inhabitants or more—who exchanged customs, languages, and ideas. Foundations of storehouses built between 500 and 350 B.C. attest to the city's abundant imports and exports. Many items, such as a Greek cup and pottery found at the site, are evidence of Ashkelon's widespread trading network.

Excavators have also unearthed Phoenician religious symbols and, among their most intriguing finds, a dog cemetery dating from the first half of the fifth century B.C. About 1,500 dogs, including puppies, which apparently died naturally, were carefully laid on their sides in a shallow pit, each with its tail wrapped around its hind legs. Stager believes that a short-lived dog cult thrived in Ashkelon. His team has also found several hoards of coins. Residents of Ashkelon began using coins during the Phoenician renaissance and eventually had one of the most active mints in Palestine. From the fourth century B.C. to the 12th century A.D. the city issued coins almost continuously. Among the most unusal coin was the mintage that served Queen Cleopatra of Egypt.

PHOENICIANS *worshiped idols and made sacrifices to them; their main god was known as Baal, or Melqart. The raw materials to make bronze—copper and tin—were shipped in ingots. Flat, four-handled ingots of copper (below) comprised much of the cargo of a proto-Phoenician ship excavated by Dr. George Bass off Cape Gelidonya, Turkey, in 1960.*

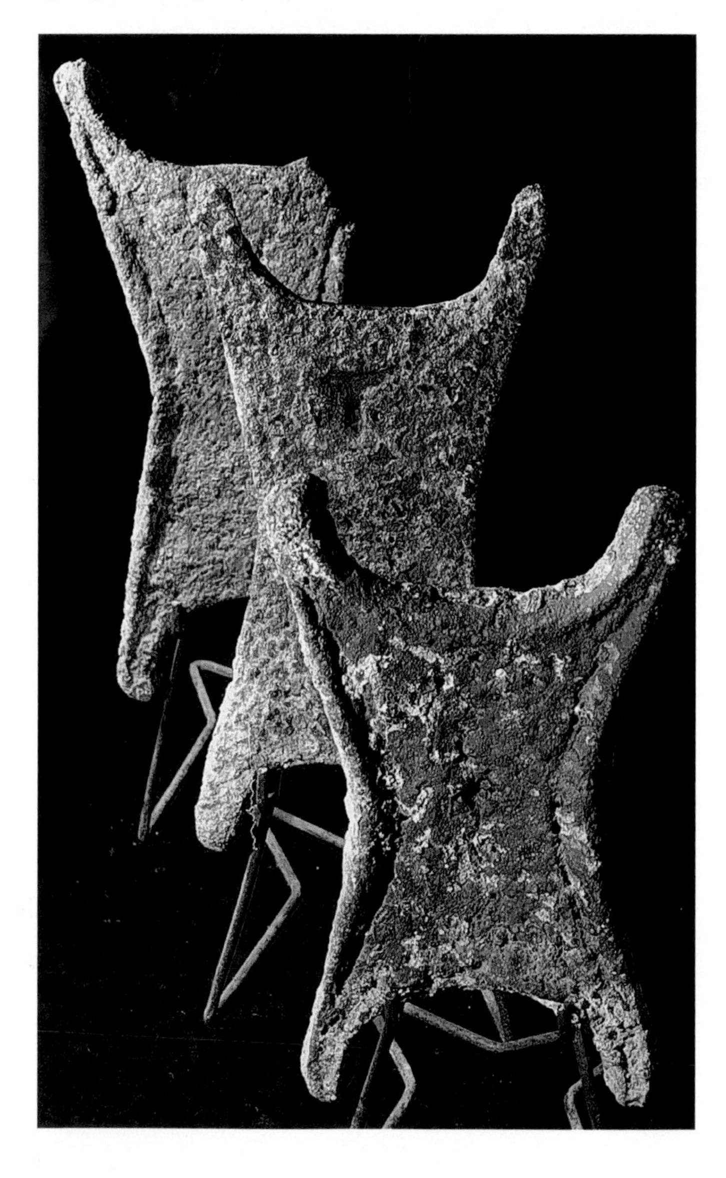

In the summer of 1997 the U.S. Navy research submarine *NR-1* videotaped what looked like shipwrecks 60 miles offshore from Ashkelon. On the seafloor 1,300 feet below the surface lay hundreds of large storage jars deposited roughly in the shape of ships. Marine geologist and oceanographer Robert Ballard and archaeologist Stager set sail in 1999 with 55 tons of equipment to study the find. After side-scan sonar located the two wrecks, the remotely controlled deepwater robot *Jason* was sent down to the site. The two ships were exposed, lying upright on the seafloor in 6-foot-deep elongated depressions. When the jars—videotaped by *NR-1*—came into view, Stager identified them as Phoenician jars of the eighth century B.C. They were used for transporting oil or wine, with a torpedo-shaped body, narrow neck, and twin handles.

The two ships dated from the Iron Age. They set sail about 770 B.C., when the Greek poet Homer is thought to have been writing his *Odyssey.* The vessels were part of a larger Phoenician fleet. They were carrying wine, probably bound for Egypt or the newly established settlement of Carthage, perhaps traveling south from Tyre en route from Ashkelon to points west. Less than two days out of their home port of Tyre, the ships apparently foundered in a storm. They hit the seafloor about two miles apart. The ships did not capsize but may have been swamped by waves. They landed on the bottom upright, with their cargoes intact, and were buried up to their decks in soft sediment; few amphorae broke in the sinking. The intense pressure, lack of sunlight, cold, and minimal sedimentation characteristic of the deep sea preserved virtually everything except the exposed wood of the ships. Wood-boring organisms consumed the upper parts, but the wooden bottoms of the ships, buried in six feet of mud, may be preserved. Over time, currents washed away the sediments, exposing the amphorae, heavy stone anchors, and some crockery used to prepare the crew's meals.

Jason retrieved some of the cargo for study and recorded more than 800 digital images of one of the wrecks. Assembled into a composite, the pictures show hundreds of five-gallon amphorae stacked 6 feet high and arrayed in the shape of a ship. Each merchantman carried more than ten tons of wine. The larger of the vessels—nicknamed *Elissa* after the mythological founder of Carthage—was about 60 feet long and contained 396 amphorae. The smaller ship—*Tanit,* named after a Phoenician mother goddess who protected seafarers—carried 385 amphorae. More amphorae may lie below these layers. The two 2,700-year old ships still hold secrets of a history we know little about. Found nearly a quarter-mile below the sea, they are the oldest wrecks discovered in the deep sea.

Phoenician prosperity in the eastern Mediterranean began to ebb when Alexander the Great of Macedon launched his campaign of conquest in 334 B.C. Phoenician city-states were ruled by or totally swallowed by Macedonian might, and Tyre was finally destroyed completely. Like previous land powers that lacked navies, Macedonia commandeered Phoenician squadrons. At the peak of Alexander's power, his empire stretched from Egypt to India. Upon his death in 323 B.C., Macedonian influence faded as those territories split into a number of kingdoms. Egypt, however, newly powerful under King Ptolemy I, destroyed Ashkelon in 295 B.C.

Nearly 150 years later, after dominating the western Mediterranean for some 300 years, Carthage, the Phoenician's greatest colony, also fell. Three conflicts between Carthage and Rome—the Punic Wars—began in 284 B.C., when the Roman Empire threatened the city-state's colonies in Sicily. Carthage's Hannibal inflicted serious damage to the Romans after leading his elephants and army over the Alps in the Second Punic War, but lost his opportunity for final victory because he lacked the might to tackle Rome itself. In 261 B.C. Rome captured a Carthaginian war galley, copied it plank for plank, and built a hundred vessels like it. That Roman fleet wiped out

REVISITING THE GELIDONYA *site with motorized scooters some 30 years after its original exploration, archaeologist George Bass and his team recovered stirrup jars (opposite) about 150 feet from the main wreckage of the proto-Phoenician ship. Phoenician ointment jars and amphorae (above) form part of a private collection in Beirut.*

Phoenician power in Africa in 146 B.C., and the Romans razed Carthage.

After more than a thousand years of daring voyages, inspired commerce, and far-flung colonization, the Phoenicians fell to more militant cultures. They had expanded the boundaries of the known world and were, wrote 19th-century British historian George Rawlinson, "the great pioneers of civilization," adventuring "where none had ever gone before." Few physical remains mark their golden age, but their heritage survives throughout the world in much more than the Western alphabet. Zeno, a Phoenician from Kition on the island of Cyprus, founded the school of Stoicism in Athens; Beirut's school of law gained world renown during Roman times. Lebanese descendants of Phoenicians are acknowledged as master entrepreneurs around the globe. As a people, the Phoenicians survive; as a civilization, they were major contributors to the modern world, where their trade and traditions are alive today.

In the summer of 2003 geneticist Spencer Wells began a multiyear study of the DNA of people alive today in areas where the Phoenicians settled in Lebanon and Syria to trace their genetic lineage. By establishing the Phoenician markers in their DNA code, he will be able to track them to Tunisia, Malta, or places as far away as Cornwall, England. This exciting study will add an important new piece to the puzzle of who the Phoenicians were. ■

BOLD PIONEER: DR. GEORGE F. BASS

From the two-person, state-of-the art submersible *Carolyn,* archaeologists on the team of Dr. George F. Bass view the remnants of a merchant ship that sank more than 2,400 years ago off the western coast of Turkey (right).

One of the founders of the study of underwater archaeology, Bass of the Institute of Nautical Archaeology (INA) based at Texas A&M University, has been delving into the depths for more than 40 years. He was the first person to excavate an ancient shipwreck in its entirety on the seabed. Each excavation involves a discipline of carefully photographing, recording, and mapping artifacts before they are disturbed, raising everything, even samples of sediment for the pollen it may contain, and spending years on the interpretation and publication of the site.

Today the scuba divers who work with Bass in the field are mostly graduate students of nautical archeology who slowly excavate everything by hand and are limited to 20-minute dives twice a day to avoid the perils of the bends. In the 1999 recovery at right, amid a veritable jungle of cables, hoses, pipes, and lines—the equipment necessary to preserve and map an area—artifacts are removed by a lifting basket while vacuums carefully remove sediment that obscures the site.

Following each expedition, experts spend years on conservation, laboratory analysis, and library research that will add new information to the history of the ancient world.

One of Bass's most exciting explorations was of a proto-Phoenician ship. A Turkish sponge diver had reported a wreck in 1982 off a Turkish promontory called Ulu Burun. When

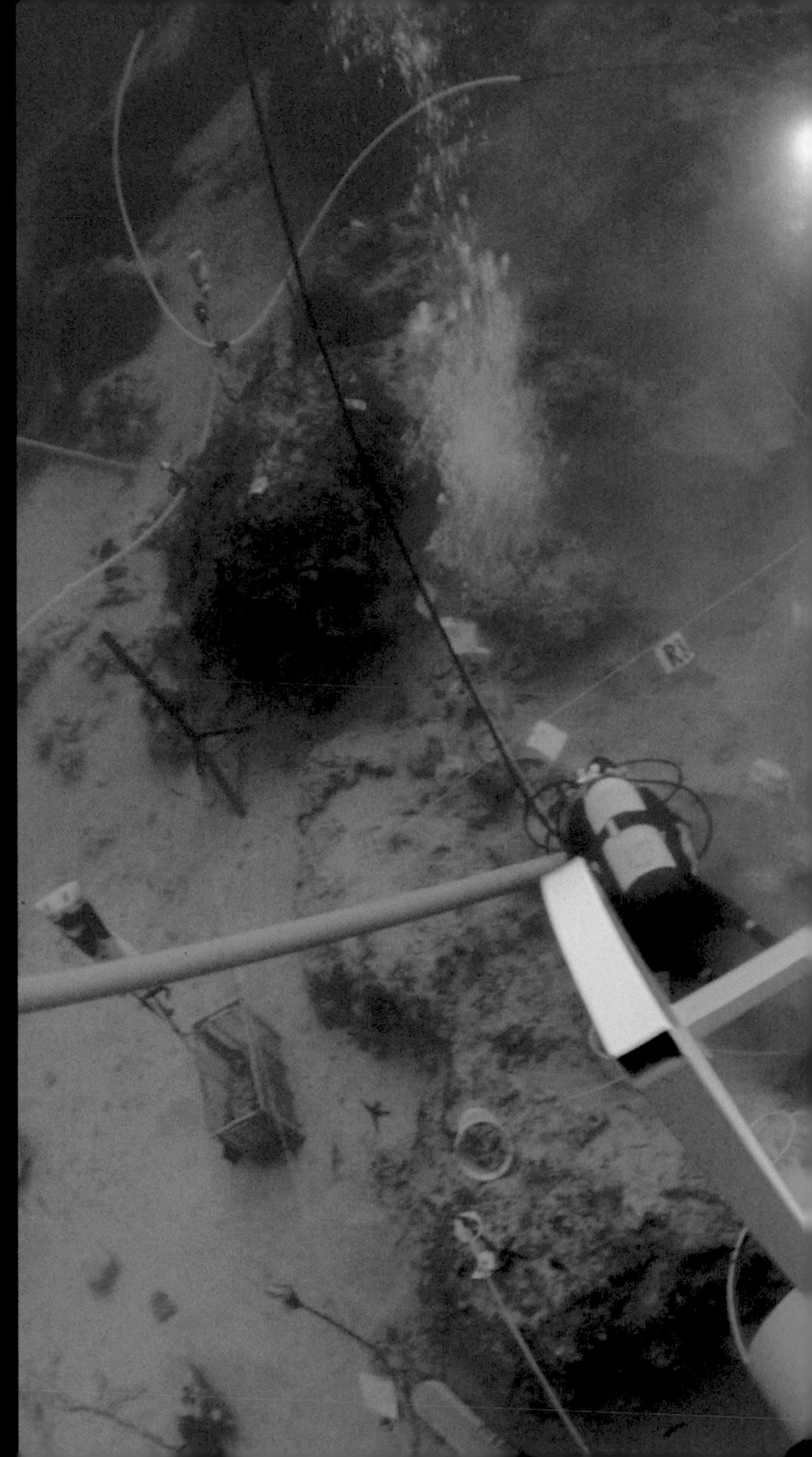

Bass and his team began to excavate in 1984, they discovered 20 tons of royal cargo. Bass with Dr. Cemal Pulak and their team from INA spent many years excavating and studying the 50-foot vessel. One extraordinary artifact (above left) lay on the wreck's western edge—a golden chalice surrounded by (top moving clockwise) a Canaanite amphora, a Canaanite flask, and a two-handled cup from Greece, called a *kylix*. Dating the artifacts with carbon-14 and from dendrochronological samples, they determined that the ship was from about 1300 B.C. and probably traveling east to west on its way to Greece or

even Egypt. The date and place of origin of the chalice, cleaned in the laboratory, still eludes archaeologists.

Based on a 14th-century B.C. Egyptian tomb painting showing the arrival of a Syrian fleet into port, Bass's team developed an idea of how the Ulu Burun ship may have looked in life above the waterline (above). The view below deck was reconstructed with the ship's recovered cargo in place, built in the shell-first manner with pegged mortise-and-tenon joints.

The illustration shows a sailor near the bow readying a *pithos,* a huge storage jar, for off-loading. Beyond him two mariners heft four handled copper ingots; to be melted with tin to make Bronze-Age tools. Near the stern a bearded Canaanite merchant and a Mycenean Greek admire a golden chalice similiar to the one excavated from the wreck. The vessel's upper features have dissolved, but some of those below the waterline have been preserved by sediment. In the stern were stored bronze swords and arrowheads, stone mace heads, ostrich egg shells, ivory, and Mycenean pottery. Amidships lay fishing nets, blue glass ingots, logs of exotic woods, and amphorae containing aromatic resin.

CHAPTER TWO

THE BLACK SEA

5400 B.C. — 63 B.C.

Frescoes of Noah and his Ark adorn the ceiling of the church of Saint-Savin-sur-Gartemps, France, recalling the great flood, which some people think took place in the Black Sea.

LURE OF THE BLACK SEA

BY ROBERT D. BALLARD

FROM CHILDHOOD I HAVE BEEN INSPIRED BY THE MYTH of Jason and the Argonauts and their perilous voyage from Greece to the distant shores of the Black Sea in search of the Golden Fleece. I dream of finding a sister ship to Jason's legendary *Argo.* In the 1970s Willard Bascom, maverick engineer and oceanographer at California's Scripps Oceanographic Institute—and one of my role models—strengthened my resolve to search the Black Sea. All the oceans of the world, really one big bucket of saltwater, contain oxygen in their deepest depths because cold, oxygen-rich water from the North and South Poles sinks like a waterfall to the bottom of the sea. Throughout the world's oceans a wood-boring mollusk called a teredo consumes oxygen as it destroys organic matter. Mother Nature is efficient at biodegrading all the uprooted trees and debris that fall into the oceans during storms. The same teredos that eat natural debris also destroy the wood of ancient ships sunk in the depths.

The Black Sea is like a bathtub without a drain. It is all but landlocked; dense salty water from the Aegean flows into it through the Dardanelles, then the Bosporus. That salty water sinks beneath the surface layer of freshwater discharged by the major rivers draining into the sea—the Danube, Dnieper,

Dniester, and Don. The lighter brackish water drains out the Bosporus while the heavier saltwater is trapped in the Black Sea depths. Filled with saltwater unable to circulate, the Black Sea has become depleted of dissolved oxygen. Bascom postulated that its oxygen-poor, or anoxic, waters, would preserve a treasure trove of ancient shipwrecks.

While the Cold War raged, researchers could not work in the Black Sea. The breakup of the Soviet Union afforded scientists a chance to research here. As I and a group of similarly dedicated scientists were in the final planning stages of a survey of the Black Sea, we read the research findings of geologists Walter Pitman and William Ryan, who had studied the Black Sea for more than 20 years and hypothesized that it was once a freshwater lake inundated in a single catastrophic flood. During the Pleistocene—Pitman and Ryan hypothesized in their book *Noah's Flood: The New Scientific Discoveries About the Event that Changed History*—the level of the freshwater lake that was the Black Sea dropped and the shoreline retreated. Then, in a few years beginning some 7,500 years ago, the level of the sea rose 550 feet above the level of the landlocked lake—a flood of such proportions that it inspired legendary tales such as the *Epic of Gilgamesh* and the story of Noah's Ark in the Bible.

In July of 1999, with the Black Sea Project, I followed in Jason's wake and sought my own golden fleece. Beamed from a sonar tethered 550 feet below the little Turkish trawler, *Guven*, digitized acoustic images danced across my computer monitor, and I contemplated an opportunity not yet afforded me in a lifetime of ocean exploration. As an explorer, I had the chance to discover an intact wooden shipwreck preserved in the oxygen-depleted waters of the Black Sea; as a scientist, I hoped to add knowledge to the legends of an ancient flood. On the expedition we found evidence of a submerged ancient freshwater lake, testimony to an influx of saltwater some 7,500 years ago. When we returned in 2000 we discovered an amazingly well preserved ancient sailing ship lying more than a thousand feet below the surface. Radiocarbon dating confirmed that a core sample our ROV *Little Hercules* retrieved from the vessel dated to A.D. 410 to 520. The first intact classical shipwreck ever found, it was also the first vessel discovered in the anoxic depths of the Black Sea. In the summer of 2003 we revisited that wreck. A new ROV, *Hercules,* retrieved wooden frames and artifacts that are now being radiocarbon-dated. From a site that may have been inhabited prior to the influx of saltwater into the Black Sea some 7,500 years ago, the ROV collected for testing sediment cores, pieces of wood, and a rectangular stone block.

Our work validated Willard Bascom's theory that the anoxic waters of the Black Sea would preserve ancient wooden wrecks; it led support to the possiblity of an ancient flood postulated in Pitman and Ryan's book. We bested Jason, recovering two golden fleeces. ■

"

AT THIS TIME, THE SEA WAS NOT NAVIGABLE AND WAS CALLED 'AXENOS' [INHOSPITABLE] BECAUSE OF ITS WINTRY STORMS AND THE FEROCITY OF THE TRIBES THAT LIVED AROUND IT, AND PARTICULARLY THE SCYTHIANS IN THAT THEY SACRIFICED STRANGERS, ... BUT LATER IT WAS CALLED 'EUXEINOS' [FRIENDLY TO STRANGERS] WHEN THE IONIANS FOUNDED CITIES ON THE SEABOARD.

"

STRABO, "GEOGRAPHY"

IFE

ROBOTIC RESCUE

A new remotely operated vehicle (ROV) Hercules, *made its spectacular debut in the summer of 2003, when Bob Ballard and his team returned to a previously explored site off Sinop, Turkey, in the Black Sea. Developed by the Institute for Exploration in Mystic, Connecticut, and the new Institute for Archaeology at the University of Rhode Island's Graduate School of Oceanography, the* Hercules *uses a high-definition camera, a subbottom profiling system, two sensitive manipulator arms, and a box of tools. On board the support ship* Knorr, *the ROV passes inspection (opposite). At the site of a 1,500-year-old wreck more than 1,000 feet below the surface,* Hercules *applies a brush to dust a beam from the ship (right bottom).* Hercules *used its manipulator to break up sediment or to scoop silt from the site. Next to the preserved beam, the ROV unearthed an amphora (right middle).* Hercules *carried its finds to an elevator lined with hollow glass floats that added buoyancy for removal to the surface (right top).*

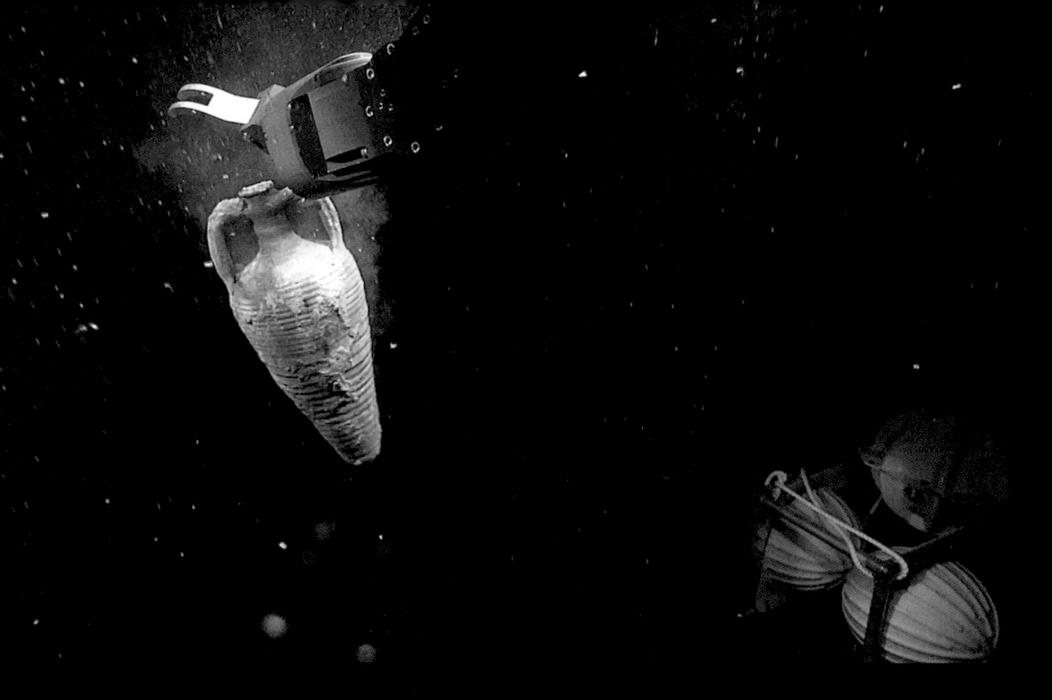

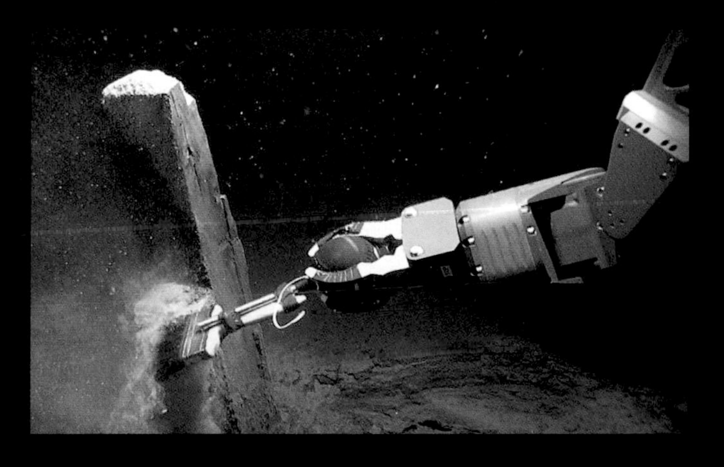

CROSSROADS OF CULTURES

MARINERS HAVE TRADED ALONG THE BLACK SEA for millennia, exchanging goods and customs between East and West. The Balkan Peninsula thrusts southward from Europe on its eastern edge, and the peninsula of Asia Minor projects westward from Asia along its southern border. The kidney-shaped Black Sea is about one-sixth the size of the Mediterranean. At its largest, it is about 750 miles long and 350 miles wide; at its narrowest, between the peninsula of the Crimea and Turkey, it stretches for 144 miles. Its only connection with the Mediterranean lies through the narrow channel formed by the Bosporus, the Sea of Marmara, and the Dardanelles Strait.

Exploration and maritime commerce began here during the Bronze Age in about 3500 B.C.; archaeological artifacts unearthed in regions surrounding the Black Sea suggest contact between European, Anatolian, and Eurasian cultures. The Don River Delta, on the eastern shore, was the northeast corner of the Mediterranean world. With the rise of Greek colonization between 800 and 700 B.C., the Black Sea became a major crossroads of the ancient world. To Greek navigators, who dreaded its ferocious storms and the hostile peoples who lived along its shores, these waters were *axenos*—inhospitable.

For centuries the area has been difficult to explore. In 1453, when the Turks captured Constantinople (present-day Istanbul), they closed the sea to all but their own ships. The Soviet Union later controlled much of the area and made it off-limits during the Cold War. With the disintegration of the Soviet Union, the Black Sea opened as an economic unit; scientists can now study it as it was in ancient times—a maritime community.

Previous archaeological evidence of ancient trade in the area has been limited largely to land. In the last 150 years some 15 ancient Bosporan towns and hundreds of rural settlements have been excavated. Amphorae and tiles that were manufactured in Turkey and unearthed at sites along the north-central coast of the Black Sea attest to north-south commerce from the central southern coast to the Crimea. Underwater excavations since the mid-1950s have revealed submerged sections of cities and small towns. In the last few years deepwater archaeology has broadened our knowledge of maritime trade in the Black Sea and added to our understanding of its geology.

Early in the first millennium B.C. nomads on horseback roamed the steppes north and east of the Black Sea. In successive waves of migrations, people streamed into the region. Mounted Cimmerian raiders haunted the southern shores. Thracians migrated into northwestern Anatolia and came to dominate the mountains and valleys of the Danube River west of the sea. Warrior-chieftains led the various tribes, which farmed, herded, raided, and traded.

From the Volga River Basin west across the grasslands of Eurasia, Scythians, as the Greeks later named them, trailed herds of sheep, cattle, and horses as they raided

SETTLERS FROM *a Greek colony at Miletus built Sinop's walls almost three thousand years ago; from this harbor on the southern shore of the Black Sea they ventured north from Turkey to establish trading posts on its opposite coast. Greeks founded some 250 city-colonies along the shores of the Black Sea between 750 and 550 B.C.*

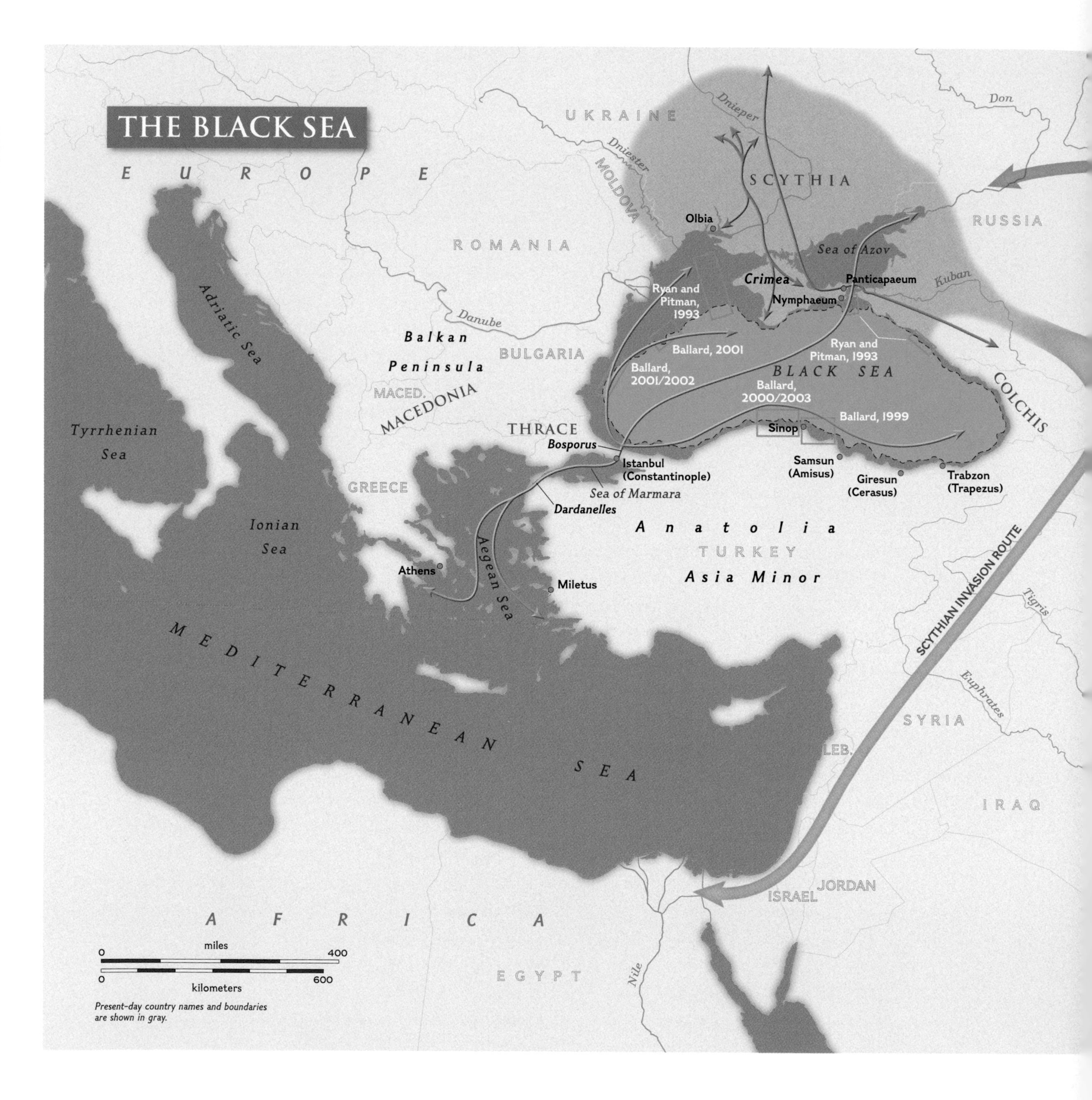

THE BLACK SEA
E U R O P E
UKRAINE
Dnieper
Don
Dniester
MOLDOVA
SCYTHIA
RUSSIA
Olbia
ROMANIA
Sea of Azov
Crimea
Panticapaeum
Kuban
Nymphaeum
Ryan and Pitman, 1993
Danube
Adriatic Sea
Balkan Peninsula
BULGARIA
Ballard, 2001
Ryan and Pitman, 1993
Ballard, 2001/2002
BLACK SEA
Ballard, 2000/2003
MACED.
MACEDONIA
COLCHIS
Ballard, 1999
Tyrrhenian Sea
THRACE
Sinop
Bosporus
Istanbul (Constantinople)
Samsun (Amisus)
Giresun (Cerasus)
Trabzon (Trapezus)
GREECE
Sea of Marmara
Dardanelles
Ionian Sea
Anatolia
TURKEY
Asia Minor
Athens
Miletus
Aegean Sea
SCYTHIAN INVASION ROUTE
Tigris
MEDITERRANEAN SEA
Euphrates
SYRIA
LEB.
IRAQ
ISRAEL
JORDAN
AFRICA
miles
0
400
0
600
kilometers
EGYPT
Nile
Present-day country names and boundaries are shown in gray.

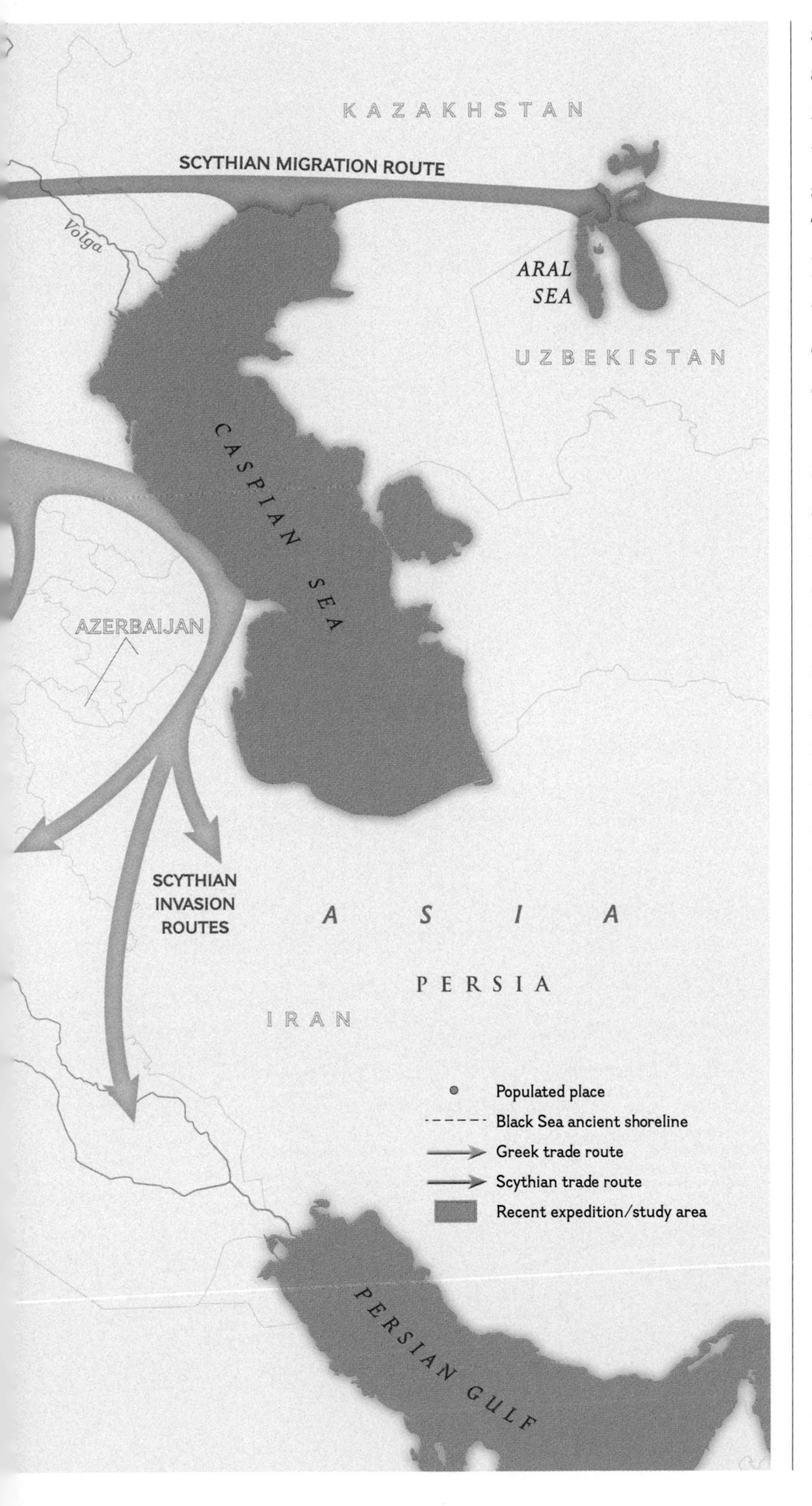

settlements in the Crimea along the northern rim of the sea. The Scythians kept no written record and, like other nomads, left few permanent traces. Kurgans, impressive timber-roof burial mounds, some several stories high, are among the few remains that mark their passing. Thousands of the sepulchers dot the landscape of the Black Sea. Some Scythians continued to raid and exact tribute from local tribes. Others settled on the steppes, establishing camps where they raised horses, sheep, and cattle and cultivated grain, which they sold or traded.

Scythians already inhabited the Crimean steppes and inland plains when Greek colonists first came ashore and set up permanent trading posts in the eighth century B.C. The Scythians made no attempt to drive the Greeks out; instead they plundered settlements, extorted tribute, and traded grain, furs, and cattle for wine, crafts, and metalwork. In some places, though, these groups lived side by side with the colonists. From the sixth to fourth centuries B.C., the Scythians ruled a confederation of tribes along the northern shore of the Black Sea and grew wealthy as middlemen between local grain producers and Greek traders. In the fifth century B.C., Greek historian Herodotus described many Scythian customs. They consumed wine outrageously, drank the blood of the first man they had killed in battle, and collected enemy scalps as trophies. Herodotus dubbed a particularly powerful and prosperous group near the Black Sea the Royal Scythians. These Scythian rulers amassed fortunes in jewelry and worked metals, and each king's wealth was buried with him in his kurgan. The first evidence of the complex burial rites and rich heritage of the nomadic Scythians was found in the early 18th century when grave robbers in Siberia began to unearth magnificent gold artifacts.

The Greek poet Homer also described the Scythians. Their horses, he noted, are "whiter than snow and swift as the sea wind."

Greek settlers along the Black Sea prospered as well. Their first foray into the area, probably in the second half of the eighth century B.C., may have inspired the legend of Jason and the Argonauts. In the beginning, Greek colonists traded in fish; later they planted grain.

SCYTHIAN *horsemen like this one depicted in an ornament from a Crimean kurgan, or burial mound, dominated most of modern-day Ukraine and the plains of southern Russia during the fourth century B.C. Another kurgan (opposite) yielded the gilded remains of a noblewoman.*

Between 750 and 550 B.C. the Greeks founded some 250 city-colonies along the Black Sea. Settlers from the Greek colony at Miletus, in present-day Turkey, established distribution centers at Sinop and Trapezus on the Black Sea's southern shore, where they traded their pottery, wine, and textiles for metals, oil, and grain. They also founded a colony at Panticapaeum on the eastern edge of the Crimean Peninsula and one named Olbia east of what is now the city of Odessa. From the sixth century B.C. on, trading vessels filled with corn and wheat from the Ukraine and northern Caucasus regularly plied the Black Sea to Greece.

From the rich fishing areas in the Sea of Azov north of the Black Sea, they carried what Romans later called *garum,* a fish sauce, and salted fish to the eastern Mediterranean. Trade from north to south developed steadily in fur, hides, and slaves as the commercial boom increased. Wine, olive oil, gold, and silver were shipped north from Greece. In return for their assistance and cooperation, local Scythian chieftains received goldwork, silverwork, weaponry, and clothing.

The Greeks remained guests on the Black Sea shores but dominated the maritime world. By 480 B.C. clashes with local Scythian inhabitants inspired some 30 Greek colonies to combine, establishing a Bosporan state. Panticapaeum grew as the capital of that kingdom in the northeastern corner of the Black Sea. It became an empire that survived as a center of wealth, industry, and power for more than 700 years.

The fifth century B.C. also saw the rise of the Achaemenid Empire of the Persians and the waning of Greek power in the eastern Mediterranean. Persian King Darius controlled the southern shores of the Black Sea and invaded Scythia across the Danube. Darius finally retreated, and the Scythians remained the major force from the Danube to the Don River throughout much of the first half of the fourth century B.C. Then another group of Iranian-speaking nomads, the Sarmatians, moved west into the Don area and pushed the Scythians west toward the Danube. The Sarmatians became the leaders of a wealthy empire that, at its widest, stretched from the Dnieper River to Colchis in the southeastern corner of the Black Sea, where Sarmatian and Scythian cultures merged for two centuries.

From his headquarters south of the sea, Philip II of Macedon expanded his empire. From 354 to 339 B.C. his army subjugated local inhabitants, including those of Thrace, until his Balkan kingdom extended from the Aegean coast to the lower Danube and from the Black Sea and the Dardanelles to the Adriatic. Virtually all of Greece fell under Macedonian control. As power passed from Athens, and Greek markets declined in importance,

Bosporan rulers encouraged other trade contacts in the Aegean. Thousands of pottery fragments attest to Thracian and Egyptian interests in the northern Black Sea. A fresco of a ship in a shrine at the ancient port of Nymphaeum dates to the third century B.C. By 170 B.C. grain was being shipped to and from the Black Sea through the Bosporus, and Egyptian scarabs and amulets were finding their way north. Ships carried marble, wine, and olive oil from Greece and returned loaded with slaves, honey, wax, textiles, and salted fish. Bosporan goldsmiths and silversmiths became adept at working the metals imported from Greece, fashioning elegant bowls, jars, and weapons decorated with fanciful animals.

In the first century B.C. Mithradates VI of Pontus in northeast Asia Minor set out from his kingdom to conquer the entire eastern shore of the Black Sea. Intent on fighting the rising power of Rome, Mithradates dragged the kingdom into a 25-year-long war that ended with his defeat by Rome's celebrated leader Pompey the Great in 63 B.C. By the end of the first century B.C. Roman fleets, many manned by seafaring peoples from the Balkans, protected the shores of the Danube and Black Sea coast. The area remained under Roman sovereignty until the Goths invaded in the fourth century A.D.

In addition to archaeologists searching its shores for this mix of cultures from the Bronze Age through the Roman Empire, the Black Sea attracts oceanographers because its depths may contain shipwrecks preserved for thousands of years. Below several hundred feet, the waters of the Black Sea are unlike any other environment on Earth. They are anoxic; they lack the dissolved oxygen necessary for life. Dense, salty water from the Aegean flows into the sea through its only outlet—the Bosporus. The saltwater sinks below the layer of brackish water on the surface that is constantly replenished by five giant rivers: the Kuban, the Don, the Dnieper, the Dniester, and the Danube. Over thousands of years organic matter carried in those rivers has extinguished life in the depths of the Black Sea. Unable to circulate, the waters of the deep have become anoxic. Although bacteria in the water naturally decompose organic matter, the inrush from the five rivers is so great that Black Sea bacteria cannot keep up with the flow. In the face of such an organic load, bacteria in the Black Sea actually strip oxygen from the water, creating in the process deadly hydrogen sulfide. The wood-boring organisms that destroy all organic matter elsewhere in the world's oceans and seas do not survive in the poisonous depths of the Black Sea.

Scripps oceanographer and pioneer explorer of the deep sea, Willard Bascom postulated 25 years ago that the hydrology of the Black Sea made it a treasure trove of preserved shipwrecks. Since then the area has opened to research and exploration, and underwater archaeologists armed with technological tools of discovery have been eager to seek the remains of sunken ships that could reveal insights into centuries of history and trade.

In the summer of 1996 Robert Ballard and a skilled interdisciplinary team of scientists, scholars, and students from the Institute of Exploration and the University of Pennsylvania began an intensive study of the area, called the Black Sea Trade Project. As the focal point of their search, the expedition chose the ancient Greek harbor of Sinop, on the modern-day Turkish coast across the Black Sea from the Crimea. Greeks from Miletus ventured north from Sinop to establish trading posts around the Black Sea beginning in the seventh century B.C. Among the settlements the colony established were Amisus (Samsun) and Cerasus (Giresun). Sinop's citizens moved on as well; Greek philosopher Diogenes, the Cynic, and King Mithradates VI were both born here.

The city's well-preserved city walls and other surviving structures have attracted archaeologists since the end of World War II. Turkish archaeologist Dr. Ekrem Akurgal conducted the first excavations, unearthing a rich cemetery

and a Hellenistic temple. Dr. Fredrik Hiebert of the University of Pennsylvania Museum of Archaeology and Anthropology and an expert on ancient commerce is the chief archaeologist for the Black Sea Trade Project. During the first two summers his team completed systematic walking surveys of the Sinop area, identifying hundreds of archaeology sites dating from Neolithic times to the Ottoman period. The land team found ample evidence of 3,000 to 4,000 years of settlement and active trade, picking up more than 4,000 shards of pottery from Greek, Roman, Byzantine, and Ottoman villages.

The Black Sea Trade Project melded land and sea exploration in a way never before attempted, combing the Sinop peninsula from mountaintop to seafloor with precise surveys and high-tech remote sensing equipment. In July 1999 expedition leader Robert Ballard, David Mindell of the Massachusetts Institute of Technology, and the rest of the water team met the land team in Sinop. The original goal of the expedition was to find shipwrecks in shallow water offshore and evidence of the ancient port, then move north into greater depths to find a trail of preserved wrecks that would mark the trade route between Sinop and the Crimean Peninsula. Hiebert and Ballard were certain the 175-mile passageway would contain a hoard of wrecks of trading vessels lost at sea.

During the final planning stages of the project, the team added an additional objective. In 1998 pioneering geologists Walter Pitman and William Ryan of Columbia University co-authored the book *Noah's Flood: The New Scientific Discoveries About the Event that Changed History.* The book postulated that stories of a deluge recorded in the Bible and other ancient literature may have been inspired by a single catastrophic event: a flood so violent that it transformed the Black Sea basin from a freshwater lake to a saltwater sea, exiling those who lived along its shores. The two geologists used core samples of sediments, bathymetry, and the dating of seashells to conclude that

BOTTOMS UP: *An ornate cup, or rhyton (above), attests to the skill of metalworkers in Thrace, on the Black Sea's west coast. A griffin (below), a mythical beast with a head and wings of an eagle and hind legs and tail of a lion, decorates a gold Thracian medallion.*

MARINERS *began braving the treacherous waves of the Black Sea to trade and explore during the Bronze Age, about 3500 B.C. By 700 B.C. the Black Sea was a major cross-*

some 7,500 years ago a flood from the Mediterranean into the Black Sea suddenly broke through an earthen dam at today's Bosporus. During the Pleistocene, they hypothesized, as ice sheets trapped much of Earth's ocean water, the level of the then freshwater Black Sea lake dropped and the shoreline retreated. As global sea levels continued to rise when Ice Age melting started around 12500 B.C., the lake remained isolated. For a while it shrank, until it finally lay some 500 feet below the level of the Mediterranean, separated only by the thin isthmus of the Bosporus Valley. The rising seawater in the Sea of Marmara cut a gorge in the northern end of the Bosporus Valley through which saltwater blasted with a force 200 times that of Niagara Falls. At a rate of some six inches a day, the level of the Black Sea rose 550 feet above the surface of the freshwater lake, displacing all in its wake. The new sea settled into a brackish upper layer and a deeper, salty layer devoid of oxygen.

Pitman and Ryan's core samples proved saltwater flooding along the Crimea, off the Black Sea's northern coast. Fascinated by their more than 20 years of research, Ballard and the rest of his team wanted to test their hypothesis that the Black Sea was once a freshwater lake inundated by a single flood of biblical proportions.

The 1999 research expedition afforded the scientists that opportunity. The full blast of the ancient flood, near the Bosporus, Ballard reasoned, would have swept away any ancient lakeshore features in its path and dumped sediments over the newly flooded bottom. East of the Bosporus and largely free of siltation, the coastal waters of Sinop, on the sea's southern shore, were a good place to search for evidence of an ancient freshwater beach. Floodwaters in this area would have risen gradually, preserving the contours of the original shore.

The team's estimates soon proved to be on target. At a depth of 550 feet some 20 miles east of Sinop, the expedition's sonar revealed the distinct curve of an ancient lakeshore. The scarp, cobblestone, berm, bands of pebbles, and sandbar on the sonar image were characteristic of a submerged shoreline near the depth that Ryan and Pitman had predicted the ancient shoreline would lie. Gusty winds and rough seas made it impossible to keep a remotely operated vehicle that could hold samples steady. Deprived of their high-tech methods, Ballard and the team resorted to an antique retrieval system. Dropping a rusty, old, iron shellfish dredge 471 feet below the choppy surface, they hauled up a sample from the ancient beach. It contained wave-rounded beach stones, polished for thousands of years on the shore of the ancient freshwater lake and submerged for millennia. Follow-up dredges brought up brown and white fresh- and saltwater shells. Radiocarbon dating of the shells at Woods Hole Oceanographic Institution verified Ryan and Pitman's theory that sea

BRONZE-AGE HEADGEAR: *A somber face decorates the back of a first-century Thracian helmet from Catalka, Bulgaria (opposite). A wide brim and grillwork helped protect the head in armor designed for a Thracian gladiator (above).*

water flooded the freshwater basin of the Black Sea some 7,500 years ago.

The freshwater shellfish species had become extinct when an influx of saltwater killed them around 5400 B.C.; the saltwater species were more recent. Between the two types of shells, the difference in dating proved to be about 7,500 years, the time when the great flood postulated by Ryan and Pitman may have occurred.

In September 2000, the Black Sea Trade Project scientists, engineers, and technicians, armed with a variety of state-of-the-art sonar and imaging vehicles, returned to Sinop. The team had two goals—to find evidence of human habitation on the former inundated lakeshore west of Sinop and, farther offshore, to discover preserved shipwrecks in the anoxic depths along ancient shipping lanes between Turkey and the Crimean Peninsula.

Northern Horizon, a British trawler, served as the support ship for two new vehicles created by Ballard's Institute for Exploration for this mission's exclusive use. The first, which would provide the precise three-dimensional imaging of sites necessary to produce digital mosaic photomaps of finds, was an optical tow sled—the *Argus*—and was connected to the *Northern Horizon* by a fiber-optic cable. *Argus,* named for an intensely observant hundred-eyed monster of Greek mythology, had three pivoting video cameras, floodlights, an electronic still camera, thrusters that enabled it to hold position above a site on the ocean floor, and scanning sonar. All were housed in a stainless-steel rectangular cage. The second Institute for Exploration vehicle, a midsize ROV called *Little Hercules,* could be attached to *Argus* by a short fiber-optic tether and would perform close-in visual site inspections and artifact removal. Fully maneuverable due to its multiple thrusters, *"Li'l Herc"* was equipped with a broadcast video camera and sophisticated lights and could be fitted with a remotely operated manipulator arm to remove artifacts and carry them to the surface. The expedition would also use deep-towed side-scan sonar equipment, the Woods Hole DSL-120. Looking for evidence of submerged ancient habitations or sunken ships, the procedure would be the same with more than 200 sonar targets. The *DSL-120* would run long, slow sweeps west to east until a promising target was found. Then the team would switch to visual mode, first employing *Argus* to video-survey the site and then closing in on it with the *Little Hercules.*

The 82nd sonar target detected in a survey about 12 miles off the coast of Sinop—a rectangular-shaped site—lay where two submerged river channels converged, an ideal location for human settlement. The team took a detailed photomosaic of the area. On September 9 *Little Hercules* motored into what looked like a site of wooden poles and squared stones half buried in silt 319 feet below the surface. The monitor on board *Northern Horizon* revealed sandstone blocks set in a distinct rectangle some 24 feet long and 12 feet wide. Then an image of wooden objects filled the screen. One, about three feet long, looked like a log hewn by tools, with one end cut and the other notched; others resembled carved tree stumps or branches. The site could well be the remains of a Stone Age house constructed of mud packed around sticks supported by beams, as was traditional along the ancient Black Sea coast. *Li'l Herc* collected wood bits from the site for radiocarbon dating but was unable to retrieve a sample from one of the logs that looked hand-hewn. The small wood samples turned out to be less than 200 years old, just trash that had drifted along the bottom. A mud sample taken from among the stone blocks of the site, however, may have been part of a wattle-and-daub wall. These were substances worked by human hands beside the ancient shoreline of the Black Sea—material that must have existed here before Mediterranean seawater flowed into the freshwater lake. Mollusk shells retrieved from the area were all saltwater species; this location had never been submerged in the

THE MYTH OF JASON AND THE ARGONAUTS

The first nautical saga in Western literature, Jason's search for the Golden Fleece was a tale "on all men's lips," as told by Homer, Aeschylus, and Socrates, long before the poet Apollonius of Rhodes penned the epic poem *Argonautica* in the third century B.C.

From his home in northern Greece, Prince Jason and his crew sailed the galley *Argo* west to far-distant Colchis, a land at the eastern shores of the Black Sea, today's Republic of Georgia. To reclaim his throne from his evil uncle, Pelias, Jason had to redeem a fleece of gold hung in an oak tree in a grove sacred to Ares and guarded by a huge and deadly serpent that never slept.

The goddess Athena helped Jason build the *Argo*—whose name meant "swift"—and properly furnish the ship for the long voyage. In the galley's bow she placed a piece of oak that had the power of

Jason escapes in triumph with the Golden Fleece in a 17th-century oil by Erasmus Quellyn.

Guided by a dove, the Argonauts row safely through the Clashing Rocks, or Symplegades, that guarded the entrance to the Black Sea.

speech. Jason and his crew of Argonauts were some of the greatest men in Greece, all heroes of renown, among them Hercules, strongest of all men, and Orpheus, most lyrical of musicians. Some, like the great horseman Castor and the fleet-footed Euphemus, were endowed with special gifts they could use to help Jason perform his heroic feats.

From Iolcus, in Thessaly, Jason and his crew sailed first to the island of Lemnos, whose maidens, the Amazons, proved so irresistible that the Argonauts spent an entire year there. In Thrace, two of Jason's men freed the country's king from his torture by Harpies, winged monsters with hooked beaks and claws that swooped down on his every meal, snatching most of the food away and fouling the remains. In gratitude the king foretold how the Argonauts' voyage would unfold and warned them of its many perils.

Near the western coast of the Black Sea, the king's advice saved them from being crushed by the Symplegades, or Clashing Rocks. No ship had ever passed between the two gigantic boulders that, driven by the force of winds, clashed together to smash ships like eggshells. The Argonauts released a dove to fly between the rocks, then rowed hard to follow the bird safely through the passage as the rocks snapped together behind them.

When the *Argo* finally landed at Colchis, its king demanded that before Jason could take the Golden Fleece he must yoke a pair of fire-breathing bulls together to plow a large field. He must plant dragon's teeth in the field, then kill all the armed warriors that would spring from them.

Medea, the king's young daughter, who fell in love with Jason, gave him a magic ointment that would temporarily protect him from harm. As fire flared from the bulls' nostrils and their breath ignited the grass, Jason stroked the angry beasts, yoked them, then cut open the soil. When warriors sprang forth ready to hurl their sharp spears at him, Jason tossed a rock among them and turned them on each other. Only then could he proceed. With a magic herbal potion, another gift from Medea, Jason drugged the monstrous serpent that guarded the Golden Fleece and took it—and the princess—home to Iolcus.

ancient freshwater lake. The site could be a Paleolithic structure that had been on a hilltop above a river valley when the area was dry land.

Two days later, while still searching for evidence of human habitation, *Little Hercules* revealed another astonishing find west of Sinop. Late on the night of September 11, the ROV broadcast views of an ancient shipwreck about 385 feet below the surface. The wreck, probably a second to fourth century A.D. Roman trading vessel, was rich with terra-cotta amphorae, small jars, and wooden beams. The position of the cargo suggests that the ship probably foundered, descended right side up, then cracked outward when it hit firm sea bottom. The amphorae lay in two dense clusters. Carrot-shaped and speckled with black, the jars were typical of those produced in Sinop. Such trading containers, fired from volcanic sands mixed with clay, have been found all over the Black Sea coastline. They were used to carry sardines, olive oil, and wine; some found with the wreck may prove to have unbroken stoppers of clay or wood and still contain residues of their original cargo. The Sinop town symbol—an eagle gripping a dolphin in its talons—was known across the sea in the region that is today's Ukraine.

That same night the team discovered a second wreck at about 330 feet a bit north of their previous find. The ship had been 45 feet wide and at least 75 feet long. In addition to a single cluster of Sinop-type amphorae, it contained scattered pieces of long and short planks, possibly part of the hull of the ancient ship. From a large oval vessel found at the site, the team dated the wreck to the fifth to seventh century A.D.

On Tuesday, September 19, *Little Hercules* verified a third ship, from the fourth to sixth century A.D., about 345 feet north of the first wreck. The ROV broadcast three piles of silt-covered amphorae, but little wood remained. Although no sampling was conducted at the sites during the 2000 expedition, the ships found by the team date from the Roman and Byzantine periods—from the second century A.D. to the seventh century A.D.—the age of the most intense trading activity between the Sinop peninsula and the West. And all three ships were found in the upper layer of the sea, in which oxygen is dissolved in the water. How, the team wondered, did wood from the ancient wrecks survive the onslaughts of wood-borers and other marine life that destroy organic matter in the sea? Local fishermen provided the answer: It depends on fluctuating levels of poisonous waters. Sometimes, the locals reported, they catch living fish; sometimes the fish are dead and the nets are black and stink of hydrogen sulfide. Russian oceanographic texts proved the fishermen correct. In this area of the Black Sea, the border between the poisonous water is not a level plane; it bulges in the middle. Waves beneath the surface transfer anoxic water into shallower depths. Free-swimming fish can escape the fluctuating level of poison water at depths of between 260 and 600 feet. Benthic organisms—ocean-bottom dwellers such as shellfish and mollusks, among them the wood-boring teredo worm—die because they cannot escape. Thus wood in the Black Sea can survive in depths shallower than the team expected, waters that are not fully anoxic.

Meeting one of the goals of the 2000 study—to determine if an ancient deep-water trade route existed between Sinop and the Crimea—proved difficult. The continental shelf off northern Turkey drops off steeply. Earthquakes have been reshaping this area on the Anatolian Fault for millions of years, and the slopes are littered with debris from landslides. The ocean floor just north of Sinop plunges precipitously from 300 feet to more than 3,000 feet deep and is a complex of massive slumps and landslides. Searches with *DSL-20, Argus* and *Little Herc* proved difficult and unproductive. East of Sinop, where the terrain within the anoxic layer is broad and more level, sonar detected several targets, but no ancient shipwrecks were found.

As the 2000 Black Sea Expedition drew to a close, the team still hoped to attain the goal of finding a preserved wreck in the anoxic layer of the sea. The *Northern Horizon* steamed north beyond the coastal shelf to explore an area between a region of sand waves and the beginning of landslides and slumping, at depths between 600 and 1,800 feet. Only two days before the project ended, *Argus* and *Little Hercules,* linked as a team 1,137 feet below the surface, crept toward a target that gleamed blue on the screen of the scanning scope. A tall pole rising from the bottom gloom appeared on the video monitor. It looked like a tapered tree trunk rising from a nest of wooden posts and timbers. Rope remnants were cross-knotted near the top. The team agreed they were looking at the 35-foot-tall mast of an ancient sailing ship. Incredibly well preserved, it sat almost upright on the soft bottom. A pair of vertical stanchions with neatly carved mortise notches stood near the mast. *Argus's* floodlights illuminated the entire 39-foot hull of the ship buried up to its deck in muddy silt. Little of the bow section was visible, but 16- to 26-foot-long deck planks were apparent beneath a thin layer of gray-brown sediment. The rudder support remained, but the steering oar was missing, suggesting the violence of the squall that sank the ship.

There were no lines or sailcloth; the storm had probably shredded them. A thick wooden spar with a rounded tenon end lay on the deck amid the timbers and hull frames it smashed when it fell. This was clearly a ship carved skillfully by hand, but over hundreds of years shipbuilding traditions along the Black Sea have changed little. The team could confirm the antiquity of the ship only when they saw no metal fittings among the remains. The fasteners visible were primitive wooden treenails—pegs driven through drilled holes.

While the crew battled blinding rain and stormy swells, *Little Hercules* retrieved a core of wood from the rudder support of the vessel. It was the only physical contact the expedition had with the wreck, but it was crucial. The wood was identified as fir, a timber traditionally used in boatbuilding in the Black Sea region. The type of wood, in addition to the boat's size and construction, suggest that it was built locally. Radiocarbon dating confirmed that it dated to the late Roman–early Byzantine period—between A.D. 410 and 520. The vessel was 1,500 years old, yet its wood looked newly hewn. The first intact classical shipwreck ever found, it is also the first vessel discovered in the anoxic depths of the Black Sea. Expeditions that revisit these waters will shed more light on Black Sea traditions, technology, and trade—and will surely find many more well-preserved wrecks.

In 2001 the Black Sea Trade Project shifted its focus from the Turkish coast to the shores of Romania and Bulgaria, in the western sea. A sonar search revealed a number of targets that might prove to be shipwrecks. Working with the Bulgarian Academy of Sciences' Institute of Oceanography in the summer of 2002, chief scientist of the expedition, Dwight Coleman, dove in a Bulgarian submersible to identify several shipwrecks. One, a vessel 275 feet below the surface several miles off the coast of Bulgaria, proved particularly important.

A pile of some 20 to 30 amphorae that looked like the type manufactured near Sinop, Turkey, lay exposed on the surface layer of the wreck—all that the team could see. The clay jars looked alike and were surprisingly large, more than three feet tall. Coleman retrieved one of them, all that the team could accomplish at the time. Radiocarbon dating indicated the wreck was ancient; it sank between the

CARVED COLUMNS *at the pavilion of Apadana portray tribute from far-flung lands sent to Persian King Darius in his ceremonial capital of Persepolis, in present-day Iran, in the fifth century B.C. Darius, who controlled the southern shores of the Black Sea, invaded Scythia but failed to add it to an empire that grew to cover nearly two million square miles.*

STONEMASONS *carved the tomb of Darius I from limestone cliffs near Persepolis (opposite); above the facade, the king worships on a dais supported by his subjects. Darius I ruled Persia from 522 to 485 B.C. Relief sculpture also covers the portico of the tomb of Xerxes, his son (below), where excavation began in 1931.*

fifth and third centuries B.C. Sediments from the jar, which contained a lot of fish bones that scientists analyzed and dated, provided more insights into Black Sea trade. The bones, four to five inches long, proved to be from a six-foot freshwater catfish; cut marks on them indicated that the fish had been butchered into chunks that would fit into the amphora. Such salted and dried fish steaks, called *tarichos,* came from the Black Sea, according to first century B.C. Greek geographer Strabo, and were a food staple for the Greek army. The Aegean Sea lacked nutrients to sustain enough seafood to feed its inhabitants.

From its cargo, archaeologist Fred Hiebert theorized that the ship started its voyage on the southern shore of the Black Sea near Sinop and sailed north to the Sea of Azov. The many major rivers that flowed into that sea made it a rich fishing ground. The ship, probably a smallish supply boat, foundered along the coast of Bulgaria as it cruised along the west coast of the Black Sea toward the Mediterranean Sea.

The 2,400-year-old wreck, oldest yet found in the Black Sea, has added knowledge of an era of trade that was previously known only through written records. It will continue to do so. An expedition already planned will return to carefully map the site, then begin collection and sampling of artifacts. Although the wreck lies above the anoxic layer and was subject to wood-borers, more amphorae and other remains most certainly lie buried beneath the first layer of jars.

Recent research has verified historical accounts. The discovery of the shoreline of an ancient freshwater lake supports Pitman and Ryan's theory of an influx of saltwater or possibly even a sudden flood into the Black Sea. Well-preserved deep-sea wrecks have proven Bascom's hypothesis that the Black Sea's anoxic waters contain undisturbed history. Future expeditions will reveal much more information about the peoples and cultures that flourished along the waters bridging East and West. ■

POISONOUS DEPTHS PRESERVE

Technological advances, including long-range sonar and remotely operated vehicles (ROVs) such as *Little Hercules* (left), help underwater archaeologists raise history from the depths. A fiber-optic cable links the ROV to its 250-foot mother ship, the *Northern Horizon.*

Little Hercules performed close-in visual inspections of sites and relayed videos to monitors on the research ship during a five-year-long Black Sea Project led by Dr. Robert D. Ballard. As they watch the videos aboard the *Northern Horizon,* Ballard shows the ROV's pilot, Craig Elder, an area worthy of further inspection (above).

The project, which began in 1996, focused around the ancient Greek port of Sinop, on the northern tip of the Turkish coast. Two factors drew the team to the area between Turkey and the Crimean Peninsula. Because the Black Sea lacks oxygen, the borers that consume organic matter in the world's other seas do not survive; perhaps the wood of ancient ships would therefore be left over. The expedition hoped to find shipwrecks preserved in the anoxic waters. In addition, the scientists hoped to verify an ancient trade route. The Black Sea was a crossroads of culture, hosting sea trade for thousands of years beginning in the seventh century B.C. The project would try to trace the trade route from Sinop north some 175 miles to the Crimea, in the present-day Ukraine.

In September 2000 *Little Hercules* broadcast views of a wreck some 385 feet below the surface. Within days the ROV located two other finds at about the same depth. Amphorae littered the seafloor, and—surprisingly for such a shallow depth, well above the anoxic layer—what looked like part of the hull remained. The poison layer of the Black Sea fluctuates, so exposed wood can survive in waters not totally anoxic.

At 1,050 feet, where poisonous hydrogen sulfide has permanently replaced oxygen, the team located a fourth wreck. It lay upright on the bottom, its mast rising 35 feet. The ship lacked metal fittings, canvas sails, and rigging, which indicated the vessel was from a

very poor or very old culture. It proved to be the first intact classical shipwreck ever found, a 45-foot vessel 1,500 years old.

Frame ends that looked as if they had been carved only yesterday rose from the bottom (above). The rudder post, hand-carved stanchions, and spars of the late Roman or early Byzantine vessel also remained intact. Working with the Bulgarian Institute of Oceanography in the summer of 2002, the team located yet another shipwreck at a depth of about 275 feet several miles off the Bulgarian sea coast.

Only a pile of amphorae remained. Some 20 to 30 jars lay exposed, with more buried underneath. Radiocarbon testing indicated that the amphorae (left) date to between the fifth and the third centuries B.C.

One amphora was recovered from the seafloor for study. On the deck of the Bulgarian research vessel *Akademik,* expedition chief scientist Dwight Coleman recognized the antiquity of the jar (opposite) and sent its contents to a laboratory for analysis.

The three-foot-tall amphora contained the remains of a freshwater catfish some six feet long that had been cut into steaks and dried. The rivers flowing into the Sea of Azov sheltered a rich supply of the fish, a food mostly used to feed the Greek army. The Aegean lacked nutrients necessary to sustain a large supply of seafood.

The ship, from the golden age of Greece, was probably a supply boat that stocked up on fish in the Sea of Azov, then sank as it sailed toward the Mediterranean on its way to deliver its shipment. The oldest shipwreck found so far in the Black Sea, it testifies to the vital role the sea played in ancient maritime commerce.

CHAPTER THREE

THE EGYPTIANS

3000 B.C. — 30 B.C.

Common craft in Egypt, a felucca, its triangular sail rigged low to the mast, plies the Nile at sunset

NILE DELTA QUEST

BY ROBERT D. BALLARD

MANY PEOPLE, AND I AM AMONG THEM, ARE FOND OF THE legend of the lost continent of Atlantis—a fabulous kingdom in the sea that in a catastrophic event sank forever beneath the waves. Few of its admirers know that the myth originated in Egypt. The Greek philosopher Plato first mentioned Atlantis in his *Dialogues* in the fourth century B.C.—the only written records that specifically refer to the island continent.

According to Plato, Solon, an Athenian poet and lawgiver, traveled to the city of Sais in the Nile Delta. There, learned and elderly priests described episodes of history that the Greeks had long forgotten. Among the most important events the priests recounted was the massive invasion of the Nile Delta by a people of a rich and powerful island empire that lay beyond the Pillars of Hercules.

The Greeks appreciated the antiquity of Egyptian civilization and the Egyptian regard for history; Plato wrote that only in Egypt was the memory of the collapse of Atlantis preserved. The legend has changed and developed over time; as yet, no evidence of the story exists from Egyptian sources.

However, history does document that in about 1200 B.C. a marauding group of invaders, called the Sea Peoples, wreaked widespread havoc in the eastern Mediterranean. The Egyptians recorded that enemies came "from the midst of the sea." The attacks on Egypt by the Sea Peoples could well have furnished Plato with the theme of maritime invasion described in his original legend of Atlantis.

Ramses III, the Egyptian pharaoh who repelled an invasion by the Sea Peoples in a fierce battle at the eastern mouth of the Nile around 1155 B.C., immortalized his victory in bas-reliefs on the walls of his funerary temple at Medinet Habu, near the Valley of the Kings.

Plato himself visited Egypt for a few months following the conversation with his traveling countryman, Solon. Perhaps the stones of Medinet Habu inspired Plato's story of Atlantis. The reliefs are elaborate and vivid: The ships of the Sea Peoples, their bows and sterns carved in the shapes of water birds, are rammed and capsized by Egypt's daring and zealous defenders.

I am convinced that many myths and legends are based on fact and were inspired by specific historical events. I would like to think the lost continent of Atlantis is one of those. But no matter, it is a great yarn. And the truth is, to discover ancient shipwrecks, marine archaeologists must search where they are likely to be found. The coast of Egypt is a logical place to look. Trade between Egypt and lands to the east was an ongoing enterprise more than 5,000 years ago. In those days Egypt was the richest country in the region. Merchants from around the

Mediterranean wanted to trade with the pharaoh's people. Not only did they bring luxury goods to Egypt's doorstep, but also necessities ranging from terebinth resin for embalming the mummies to ship's timbers. Besides importing cedar logs for shipbuilding, entire newly built ships were delivered to Egypt. In one account, the ruler from Alashia, Cyprus, writes to a pharaoh, "All that thou desirest, I will bring bring up to thee.... Ships in quantity I will build."

The oldest port of the richest person in the world was located in the eastern branch of the Nile Delta: The pharaoh had all the money and all the power; whatever it was, he wanted it, and it had to come by boat. It stands to reason that some of those trading ships foundered along the way, and I am going out there to look for them.

Other explorers have already searched Egypt's coast. In the 1990s French archaeologist Jean-Yves Empereur discovered Greek and Roman wrecks dating from the fourth century B.C. to the seventh century A.D., littering the port of Alexandria, west of the Nile Delta.

Each new find will provide information about the mariners who transported the discovered objects and the civilization that built it. In the depths of the eastern Nile Delta may lie some of the carved ships of the Sea Peoples and other ancient seafarers, waiting to add new information to the legend of the lost continent and other histories. ■

“

I SAILED DOWNSTREAM
TO BEAT BACK THE ASIATICS,
WITH MY VALIANT ARMY
GOING BEFORE ME LIKE
A FLAME OF FIRE.

”

PHARAOH KAMOSE, 1570 B.C.

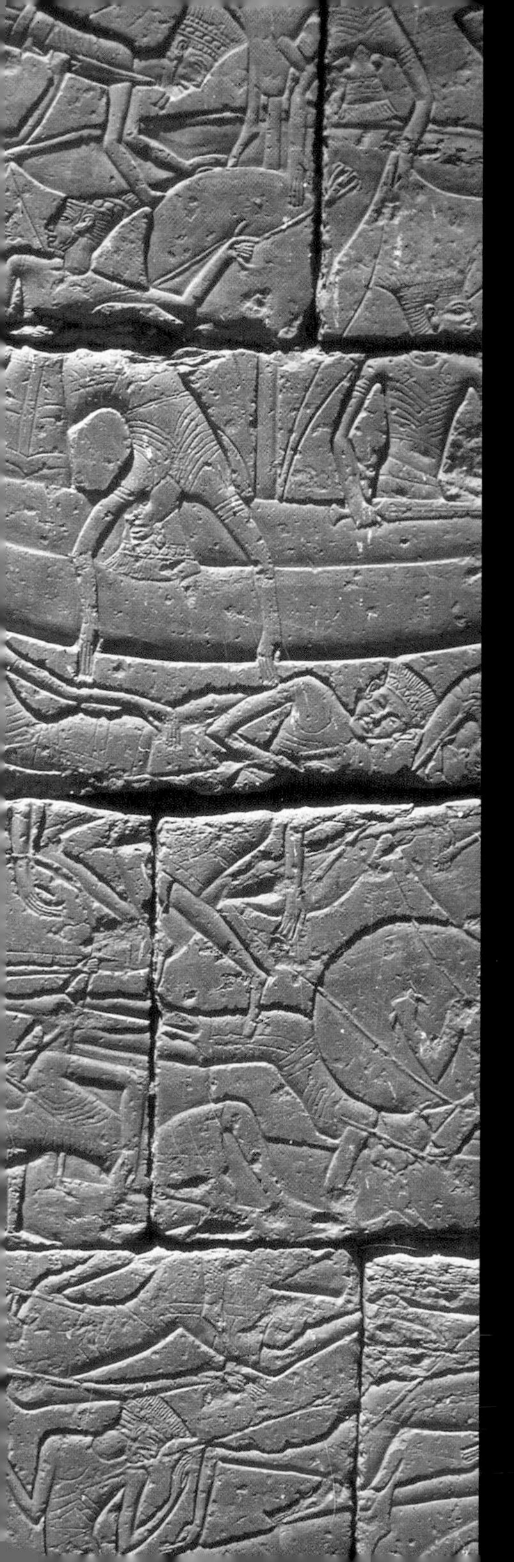

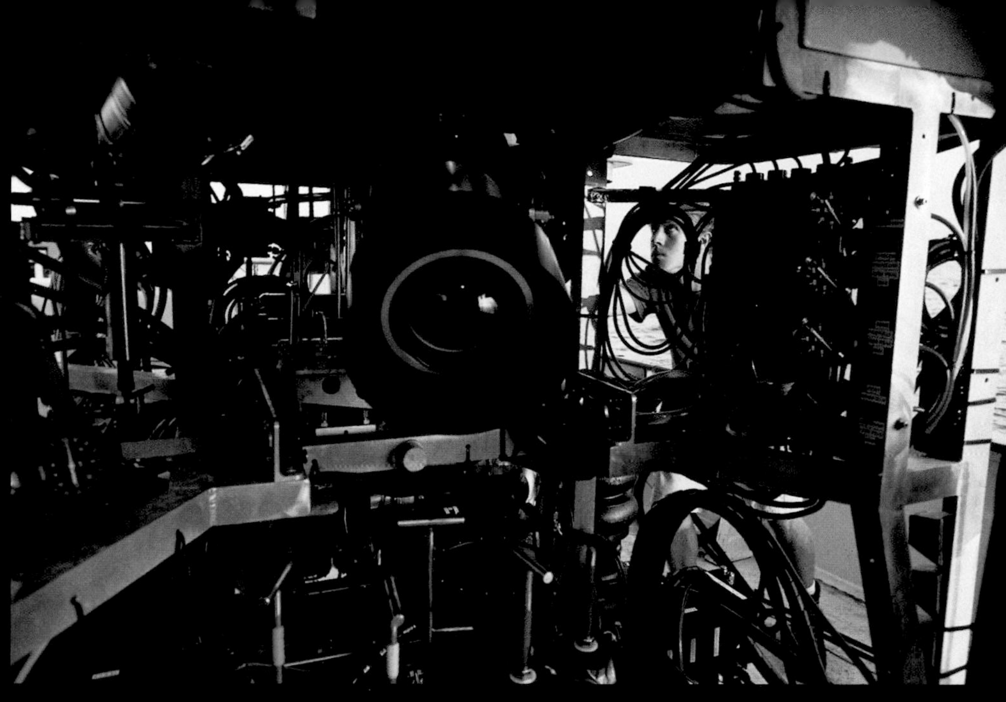

PAST AND PRESENT

Carvings on the walls of the temple of Ramses III at Medinet Habu, across from Thebes, celebrate his daring victory in the Nile Delta to repel the Sea Peoples, aggressive marauders who invaded the Mediterranean about 1200 B.C. (left). Some experts believe the invaders were Canaanites, or early Phoenicians. Robert Ballard hoped to search for remains of these war vessels and further excavate two Phoenician ships off the coast of Israel in the summer of 2003. The state-of-the art equipment for this expedition, the remotely operated vehicle (ROV) Hercules *and* Argus, *the platform to which it was tethered, were both fitted with lights and high-definition camera systems. Todd Gregory studies the giant camera eye he invented for* Hercules *(above). Unfortunately, Egyptian authorities rescinded the expedition's permit. Ballard plans to return. Combined with the superb lighting provided by these vehicles in the Mediterranean and the Black Sea,* Hercules *produced the most spectacular underwater images of wrecks made to date.*

ASSURING *a happy afterlife, a painting in the tomb of the chief physician to Amenhotep II, Nebamun, immortalizes him, his wife, and children hunting waterfowl, one of the wordly pleasures the family enjoyed. Egyptians believed such decorations, combined with luxuries used in life, ensured a happy and plentiful existence—after death.*

FROM ANCIENT EGYPT TO ALEXANDRIA

FROM ABOUT 3100 B.C. TO 343 B.C., 31 DYNASTIES of Egyptian rulers shaped one of the oldest and longest-lived civilizations on Earth. Egypt's cultural legacies are indelible, among them the great pyramids and temples, elaborate mummification, and golden treasures. Egypt began on the banks of the Nile as an agrarian society and grew to become a trading power in the Mediterranean. The Nile, broad, deep, and easily navigable for 500 miles north from the first cataract to its delta north of Cairo, was the physical and figurative heart of Egypt. The river unified and centralized its people and its culture. All travel was by boat. As early as 3500 B.C. the Egyptians used the sail, the first people in recorded history to do so, floating downriver on the current and sailing upstream with the prevailing winds. The deserts hemming the Nile Valley and dense marshes along its delta formed natural barriers that kept invaders—for the most part—at bay as Egyptian civilization developed.

An alphabet of 24 symbols, all consonants or semiconsonants, formed the basis of Egyptian writing; vowel sounds could not be expressed. Hieroglyphic signs represented different phonetic combinations. By the time Egyptian dynastic history began, when the provinces of Lower Egypt in the north and Upper Egypt in the south were united under a single god-king, traditionally known as Narmer or Menes, scribes began using hieroglyphs to record dates, events, and the names of rulers.

Narmer established his capital at Memphis, south of the Nile Delta. During the Early Dynastic Period, from about 3100 to 2686 B.C., Egyptians were no great traders; they ventured forth only in search of luxury items. Missions to Phoenician Byblos, in present-day Lebanon, brought cedar wood and olive oil to the banks of the Nile. Syrian-style pottery excavated from tombs attests to well-established trade with that area. Egyptians also sailed to Nubia, south of the first cataract of the Nile, for conquest and to trade for slaves, copper, gold, ivory, ebony, and exotic animals.

The flowering of Egyptian civilization occurred over three long periods, known as the Old, Middle, and New Kingdoms, intersected by times of social chaos and upheavals known as Intermediate Periods. It was during the Old Kingdom—2686 to 2160 B.C.—that kings built their monumental pyramids. By this time Egyptian rulers claimed divine support for their authority and enjoyed absolute power based on the belief that only they had a right to the afterlife, which would be spent with the sun god Re; ordinary folk could look forward only to a shadowy existence.

The kings, whose tombs grew more and more opulent, showered favors and wealth on craftsmen. Egyptian art was not frivolous; it was functional, usually commissioned for monuments and tombs. The common Egyptian performed the backbreaking labor of raising the massive memorials. The farming year was always dependent on the annual flooding of the Nile, but whenever the harvest ended, Egyptian farmers were conscripted to labor on the huge stone edifices.

Egyptians built their homes—from simple huts to crowded multistoried houses, spacious country estates, and royal palaces—of sun-dried mud brick. Temples and tombs, meant to last for eternity, were constructed or carved from stone. Limestone and sandstone could be cut with copper chisels. Granite, also popular, was more arduous to quarry and more difficult to transport; blocks of granite could

INLAID *with lapiz lazuli and other semiprecious stones, a portrait of Pharaoh Tut-ankhamun adorns his golden coffin. He holds the symbols of kingship—the crook employed by shepherds crossed with the flail used to thresh grain.*

weigh hundreds of tons and in some cases over a thousand. The awesome size and weight of the stone monuments still convey a sense of power. Their massive and mysterious aura is enhanced by the lack of light from windows.

Egyptian pyramid-building culminated in the Great Pyramid of Khufu (or Cheops) and its companions for Khafre and Menkaure on a limestone plateau at Giza, west of the royal capital at Memphis. Of the Seven Wonders of the Ancient World listed by Greek historian Herodotus in the fifth century B.C., only the Great Pyramid at Giza remains, rising more than 480 feet from flatlands near the Nile. An outcrop of limestone at the edge of the plateau was shaped into the Great Sphinx, with the head of a king and the body of a lion, to guard the gates of the underworld. Although the question of how the huge pyramids were built remains still unclear, the monuments attest to the logistical and engineering skills of the early Egyptians.

For centuries the Egyptians sought no lands or power beyond their borders. They kept up links with Byblos and neighboring areas. Other trade routes stretched across the Sinai desert in search of copper, malachite, and precious stones. Reliefs on the tomb of Pharaoh Sahure, who ruled about 2470 B.C., attest to his sending a fleet from Egypt to the eastern shores of the Mediterranean. A picture of the arrival of his fleet back in Egypt decorates his mortuary temple and remains the oldest clear representation of Egyptian seagoing ships.

Egypt continued to trade with Nubia to the south. One official bragged that he had retrieved 300 donkey loads of luxuries, among them incense, ebony, elephant tusks, and panther skins. At an Egyptian colony at Buhen, near the first cataract of the Nile, archaeologists have unearthed the remains of a copper smelter operated by Egyptian prospectors. Trading ebbed, however, as the kings lost control, and the Old Kingdom faded. Egypt slipped into a state of anarchy, termed the First Intermediate Period.

In about 2060 B.C. a Theban king, Mentuhotep I, reunited Upper and Lower Egypt and restored order. His rise marked the beginning of the second major period in Egyptian history, the Middle Kingdom, a 400-year interval of peace and prosperity. Amun, the patron god of Thebes, became linked with the sun god Re to become a national deity—Amun-Re. The god of the dead and resurrection, Osiris, also grew in prominence, reflecting a change in attitude. During the Middle Kingdom, the common citizen, who had no connection to a king, could gain access to the afterlife. Death was a pathway to an afterlife.

Life for the Egyptian commoner, though, was harsh and unyielding. The large population of farmers along the Nile could look forward to painful punishment if they failed to pay the annual rent and taxes on their crops at the end of each harvest—all of which went into the pharaoh's coffers or to the temples, which continued to gain land through time. The average Egyptian relied on magic, superstition, and ritual to get through each day, and his existence was short; the life expectancy of the general population was 29. The rich—some 5 percent of the population—enjoyed a sumptuous lifestyle of imported luxuries and country villas that extended their average life span to the age of 35 or 40.

The Egyptians believed in resurrection; therefore whatever their rank, they were obsessed by death and the afterlife. The wealthy were buried with the items they might need in their eternal life, and its quality depended on the provisions available. Tomb paintings ensuring a life of plenty showed tables laden with food, other furnishings, and favorite pieces of jewelry—an indicator of rank—as part of the usual funerary equipment. Even boats, the main mode of transport in life, were thought necessary for the afterlife to travel through the underworld. Much of Egypt's wealth was diverted into catering to its fascination with and preparation for death. Burial and mummification were rituals designed to preserve the integrity of the dead.

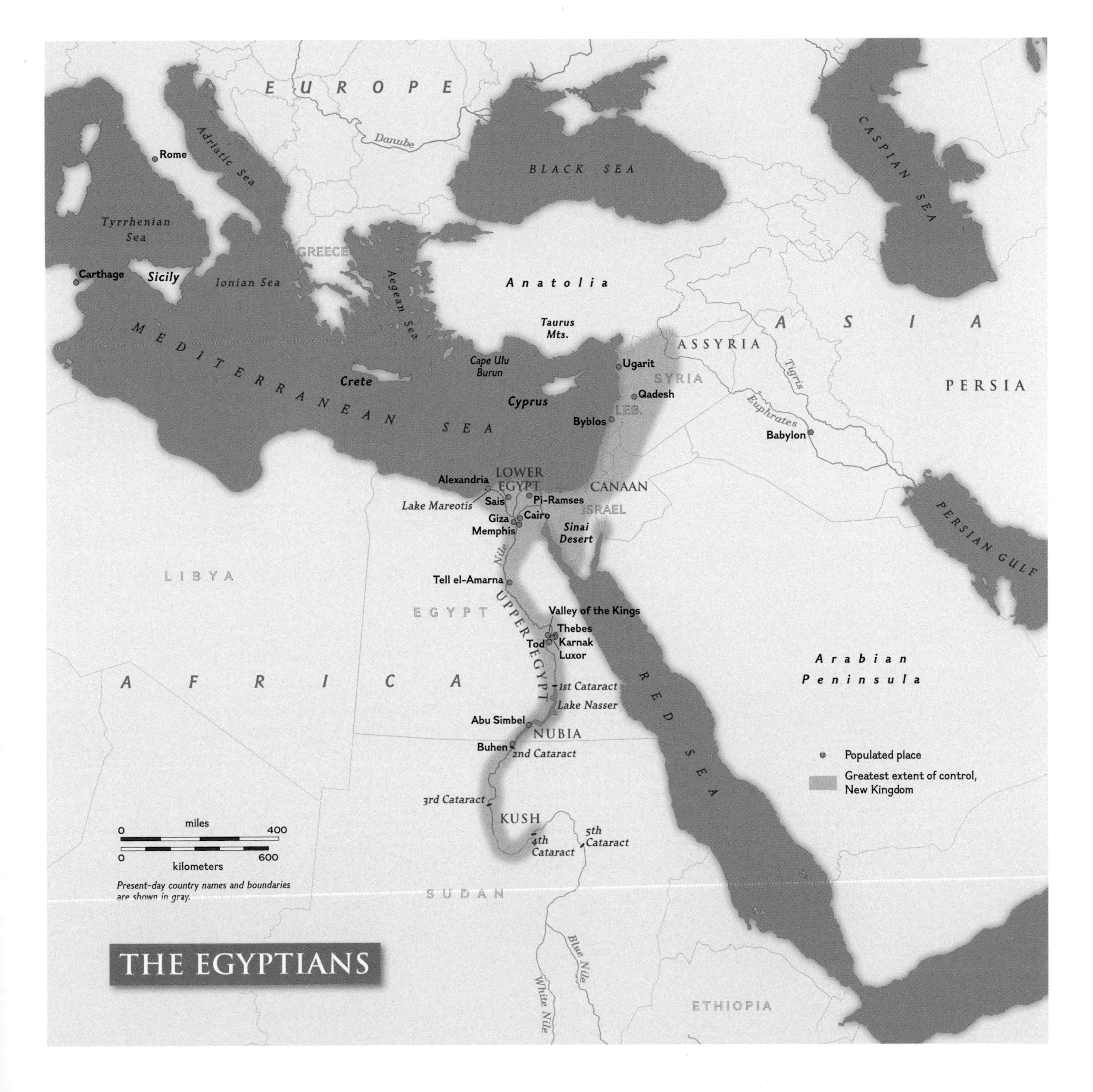
THE EGYPTIANS
EUROPE
Danube
BLACK SEA
CASPIAN SEA
Rome
Adriatic Sea
Tyrrhenian Sea
GREECE
Carthage
Sicily
Ionian Sea
Aegean Sea
Anatolia
Taurus Mts.
ASIA
ASSYRIA
MEDITERRANEAN SEA
Cape Ulu Burun
Crete
Cyprus
Ugarit
SYRIA
Qadesh
LEB.
Byblos
Tigris
Euphrates
PERSIA
Babylon
Alexandria
LOWER EGYPT
CANAAN
Sais
Pi-Ramses
Lake Mareotis
ISRAEL
Giza
Cairo
Memphis
Sinai Desert
PERSIAN GULF
LIBYA
Nile
Tell el-Amarna
EGYPT
UPPER EGYPT
Valley of the Kings
Thebes
Tod
Karnak
Luxor
Arabian Peninsula
AFRICA
1st Cataract
Lake Nasser
RED SEA
Abu Simbel
NUBIA
Buhen
2nd Cataract
Populated place
Greatest extent of control, New Kingdom
3rd Cataract
KUSH
5th Cataract
4th Cataract
miles
0
400
0
600
kilometers
Present-day country names and boundaries are shown in gray.
SUDAN
Blue Nile
White Nile
ETHIOPIA

FABRIC AND FIBER *ranked among Egypt's main exports. A tunic (left), one of the oldest pieces of intact clothing ever found, reflects a style popular for 1,500 years. A painting (above) on papyrus captures a river boat; Egyptians made paper by cutting the pith of the papyrus stalks (top) into thin strips and pressing it.*

During the Middle Kingdom, commercial activity revived, and well-worn trade routes reopened. Expeditions to Syria and Canaan began anew. About 2000 B.C. Pharaoh Sankhare Mentuhotep III sent a large expedition to the land of Punt, on the coast of modern-day Ethiopia and eastern Sudan. In a perilous journey, Egyptians built ships on the Nile, disassembled them, and carried them across the eastern desert to the Red Sea for the sea voyage.

From Nubia and, farther south, the land of Kush, gold was shipped north on the Nile to the Mediterranean. Senwosret III, who ruled from about 1878 to 1841 B.C., built and provisioned a series of forts along the Nile to safeguard trade in Nubian gold, copper, and amethysts.

Egyptian exports reached Ugarit in modern-day Syria and Byblos, and from those ports moved on to Crete and Cyprus. Middle Kingdom tombs have also yielded pieces of Minoan pottery. The court at Byblos came under strong Egyptian influence during the Middle Kingdom, and artifacts bearing the names of several Middle Kingdom pharaohs were found in the royal tombs there. Egyptian gold, linen, papyrus, leather goods, and grains were exchanged for wood, wines, oils, resins, copper, and silver. A hoard of gold and silver containers and seals excavated at Tod, in Upper Egypt, may have been made either by Minoan artisans or were inspired by them. Craftsmanship from the lands of the eastern Mediterranean, especially jewelry found in the tombs of royal princesses, reached new heights.

In the 1700s B.C. chaos again swept Egypt during the Second Intermediate Period. Invaders into the delta from the east, the Hyksos, brought confusion and instability to Egypt. The determined Hyksos, with their horses and chariots, conquered the capital of Memphis in 1663 B.C. and ruled northern Egypt for more than a hundred years. Although the Hyksos continued to control some parts of Egypt, certain nobles and ruling families carved out territories of their own and maintained relatively stable states.

LIGHT BURNISHES *the sides of three of the pyramids at Giza (above). The massive monuments are the last survivors of the Seven Wonders of the Ancient World. Seemingly the largest, the pyramid of Pharaoh Khafre stands on a rise between those of Menkaure and Khufu. The even larger Great Pyramid of Khufu rises to 482 feet. In its day, the pyramid complex was a sprawling community (right). Without benefit of wheel or pulley, the pharaohs began erecting the monuments during the Old Kingdom, some 4,500 years ago.*

GUARDIAN *of the desert west of Giza, the 240-foot Sphinx has the body of a lion and the head of a man. Masons of Pharaoh Khafre constructed the sandstone figure some 4,500 years ago, perhaps to watch over his nearby pyramid.*

SEALED IN *a stone crypt at the foot of the Great Pyramid, a ship built for King Khufu (above) was disassembled into more than a thousand parts, to be used in his afterlife. Re-Harakhty, a composite of the sun god, Re, and Horus, the god of all living things, is ferried across the heavens in a sacred papyrus raft (below).*

One of those arose in Thebes, a Nile town that had been a staging post for trade with Nubia.

About 1555 B.C. Theban princes ousted the Hyksos and established the third great period in ancient Egypt's history, the New Kingdom. Thebes became the undisputed capital, and from it New Kingdom pharaohs ruled a reunited Egypt. New Kingdom rulers, unlike their predecessors, whose splendid monuments invited robbers, opted for tombs that were well hidden. In the Valley of the Kings, an isolated dry river bed, west of the Nile and across from Thebes, they cut a royal necropolis out of the limestone cliffs. Deir-el-Medina, the village for the workers who toiled on the royal burial complexes, was a closed community; its 1,200 inhabitants were locked inside the walls each night to preserve the secrecy of the site.

At Luxor and Karnak, in ancient Thebes, the kings ordered the construction of splendid temples, obelisks, and inscriptions to honor the god Amun-Re. Karnak, which originated during the Middle Kingdom as a simple temple to the local god Amun, became the richest and most powerful temple in ancient Egypt. It was expanded and renovated for 2,000 years, eventually sprawling across 250 acres and containing 20 temples. The main temple remains one of the best preserved archaeological sites in Egypt.

The New Kingdom's outlook toward trade and conquest became more aggressive. Like their predecessors, the pharaohs presented themselves as semidivine protectors. However, while Egyptian kings had for centuries been content to simply defend their land against attack, New Kingdom rulers began to embark on military campaigns into Syria, Canaan, and south into Nubia.

Stung by the Hyksos invasion, they believed the only way to stop future incursions and restore security was to expand Egyptian power. Advancing steadily, Egypt gained control of much of the Syro-Canaanite coast. The New Kingdom pharaohs, relying on professional soldiers and using technology such as the horse-drawn chariot and the composite bow introduced by the Hyksos, developed Egypt's first professional standing army.

In the 15th century B.C. the military campaigns of

Thutmosis I subjugated Nubia in the south and also extended the Egyptian empire eastward to the Euphrates River. Under his successor, Thutmosis II, plunder from wars and tribute from foreign countries—gold, silver, lapis lazuli, and other riches—poured into the pharaoh's coffers, increasing Egypt's wealth and stimulating its culture. On his death, his wife Hatshepsut, the daughter of Thutmosis I, seized power. For the first time Egypt had a ruling queen. Reliefs on her tomb in the Valley of the Kings immortalize an expedition Hatshepsut sent from the Red Sea to Punt. Galleys with a single mast and 15 rowers on each side carried daggers, hatchets, and necklaces to barter for ebony, myrrh, gold, and ivory.

Thutmosis III, who was determined to extend Egyptian authority in Asia, rose to power when Hatshepsut died. He stockpiled Syro-Canaanite ports as forward supply bases for his army and transported his troops north by galleys during his yearly campaigns eastward. During one expedition Thutmosis III took his army across the Euphrates River, bringing Egypt into direct contact with the remote and warlike Hittites and Mitannians, and thus enlarging the Egyptian empire to its maximum extent. At its height, about 1400 B.C., the Egyptian empire stretched for more than a thousand miles.

Even in the heyday of its trade, Egypt did not sail beyond the Red Sea and the southeastern corner of the Mediterranean. Those long-distance explorations fell to civilizations of lesser power and more daring mariners. Egypt never became a maritime power; its fleets were river craft and a few merchant ships, and vessels of neighboring countries regularly carried Egyptian goods. The proto-Phoenician ship discovered near Ulu Burun, Turkey, by archaeologist George Bass and excavated by him and Cemal Pulak of Texas A&M University's Institute of Nautical Archaeology might well have been on its way to Egypt when it sank around 1300 B.C. Its goods of copper, tin, glass, terebinth resin, ivory, scrap gold, seals, and many other commodities were similar to the kinds of tribute foreigners offered to Egyptian pharaohs.

Through its trading network, however, Egypt had contact with many Mediterranean nations—Cyprus, Crete,

EVERYDAY TASKS: *Herdsmen tended Egyptian cattle (right); only the wealthy could afford the animals, which were expensive to feed and heavily taxed. Wooden models found in a tomb represent Nubian bowmen (above), who served as mercenaries for the pharaoh.*

and mainland Greece. Wealthy Egyptians enjoyed the luxuries afforded by widespread maritime commerce. They wore fine cloth woven and dyed in Phoenician cities and used objects made of copper imported from Cyprus; they built their homes of Lebanese cedar, decorated them with African ivory, and ate beef raised on the Syro-Canaanite coast. Egyptian ports flourished. Memphis attracted so many traders from abroad that a foreign quarter grew up. Ships simply pulled up on the beach to unload cargoes under the watchful eyes of customs officials who ensured that the pharaohs' tariffs were paid immediately.

Its neighbors accepted Egypt as a great power, and its pharaohs ruled supreme until the time of the 18th dynasty, under King Amenhotep IV. Bent on completely eradicat-

STILL WATERS *of a sacred lake mirror the Temple of Amun-Re in Karnak, near Thebes, capital of the New Kingdom. Dedicated to the god of Thebes melded with the sun god, the temple became the richest in ancient Egypt.*

ing the old gods, the unconventional Amenhotep renamed himself Akhenaten, or "Servant of the Aten," the radiant god of the sun disk. About 1350 B.C. he established a new capital at a site known now as Tell el-Amarna, and he forcibly replaced the traditional polytheistic religion with a new one, recognizing only a single sun god, the Aten, as the universal deity. A temple to him dominated his new city; untraditional in the extreme, it was open to the sun, contained no shrine or cult statue, and was not served by priests. Perhaps to divert attention from Theban priests to himself, Akhenaten revived the old belief that the king was the only intermediary between the people and their god. His wife, the legendary beauty Queen Nefertiti, played a prominent role in the new religious ceremonies.

El-Amarna was spacious and gracious. The city center contained government offices and lesser temples; a business district stretched northward; a garden-filled residential area of single-story houses formed the southern boundary. Within the 17-year-reign of Akhenaten, el-Amarna grew to shelter 20,000 to 50,000 inhabitants. But its glory was short-lived.

Akhenaten had closed many temples in an attempt to destroy rival gods and had confiscated their wealth for his own. His attempt to transform the spiritual life of Egypt, though unsuccessful, left the country spent and in disarray. After his death el-Amarna was abandoned, and the capital moved back to Memphis. Theban priests razed his temples and palace and tried to restore the old order under the boy-king Tutankhamun.

Howard Carter and Lord Carnarvon's 1922 discovery of Tutankhamun's tomb, the only one in the Valley of the Kings known to have survived with its treasures intact, made the obscure King Tut famous. In reality, the young king died too soon, at age 18, to resolve the upheavals that were tearing Egypt apart. And he left no male heir.

Power passed to the military. Ramses I, an elderly general who had no proper claim to the throne, eventually

BURIAL GOLD *accompanied King Tut to his tomb. A dagger tucked into a band at his waist glitters with glass and semiprecious stones. The goddess Nekhet, shown as a vulture, centers a flexible collar that covered the chest of the mummified young king.*

established himself as king and fought to maintain the empire. His grandson, Ramses II, marched north to battle the growing power of the Hittites. Proclamations on the walls of Theban temples attribute the battle of Qadesh as a great victory for Ramses II; Hittite accounts aver that the Egyptian army was surrounded and managed only to escape. Recognizing that their resources were limited and that other enemies threatened, Egypt and the Hittites made history's first recorded peace treaty in 1268 B.C.

During the 50 years that this peace lasted, Ramses II indulged his obsession with building, erecting palaces, temples, statues, and festival halls in his name. Nearly half of the ancient temples that survive in Egypt were created by him. At Karnak, Ramses built a 12-acre columned hall. His tomb was the most opulent in the Valley of the Kings, and east of it he had constructed the Ramasseum, a massive mortuary temple. At Abu Simbel, deep in Nubia near the second cataract, he established a site intended to awe his far-flung subjects. There, a colossal temple guarded by four seated statues of Ramses himself honored the king as a god. The statues are 67 feet tall; just the head of one weighs more than 20 tons. A second temple was dedicated to his favorite queen.

Today Abu Simbel is a triumph of conservation. As the waters of Lake Nasser rose with the completion of the Aswan Dam to control flooding of the Nile, the entire site was removed, stone by stone, lifted some 200 feet up a hill, and painstakingly reassembled. The project was completed in 1968 to save the temple from drowning. To this day, it rivals the pyramids as one of the most popular tourist destinations in Egypt.

Ramses II moved the Egyptian capital north from Thebes to the eastern delta at Pi-Ramses, a location designed to help him maintain control of Asia. But the pharaohs who succeeded Ramses faced unsettled times throughout the eastern Mediterranean and western Asia. About 1200 B.C. marauders known collectively as the Sea Peoples began to wreak havoc along the coasts from Asia Minor to Egypt.

At the edge of the western delta, King Merenptah repulsed a confederacy of Libyan tribes and seaborne troops intent on settling in Egypt. Early in the reign of Ramses III (1190 to 1158 B.C.), Libyan forces attacked again in the delta.

Three years later the Sea Peoples tried by land and sea to invade and occupy Egypt. A flotilla of their ships reached the delta, probably near the mouth of the easternmost branch of the Nile, where the forces of Ramses III fought a desperate battle to repel the assault. The Egyptians used their ships as floating fire bases. With the notable exception of the grapnel, which they used to capsize opposing ships, they employed no specifically nautical weapons.

Carvings on a temple wall at Medinet Habu, near the Valley of the Kings, immortalize the encounter. These once-painted reliefs are the earliest recorded depiction of a naval battle and the best illustration extant of Egyptian warships in action. The Egyptian ships had high bulwarks, curved hulls, and low prows and sterns; these vessels could be propelled by oars or sail. The galleys of the Sea Peoples were angular, the bow and stern posts ending in carvings of the heads of water birds. A scribe of Ramses III left a dramatic written account of the encounter; it is the first naval battle of which a contemporary description has survived. The bas-reliefs and the account both celebrate Ramses III's daring offensive. The Egyptians pounced on the invaders; they loosed a barrage of arrows and slung stones at them, rammed and capsized their boats, then closed in on them with spears and battle maces.

The Egyptians repelled the Sea Peoples on land as well, attacking them from horse-drawn chariots introduced to Egypt by the early Hyksos invaders. No evidence of the battle has been recovered, but many underwater archaeologists believe it exists. Oceanographer Robert Ballard hopes to find remains from the land battle as well as sunken vessels on his next expedition to the Nile Delta.

THE TALE OF THE SHIPWRECKED SAILOR

Boatmen sail the Nile in a mural in the tomb of New Kingdom Pharaoh Sennefer in the Valley of the Nobles outside Thebes.

More than four thousand years ago, during the reign of Amenemhet II, a pharaoh of the 12th dynasty of Egypt's Middle Kingdom, a travel-weary wanderer meandered through the busy streets of the Royal Court in Thebes. Seeking an audience with the pharaoh, the sailor accosted his right-hand man, the grand vizier, and shared his amazing story with him.

Sent by the pharaoh, the traveler had sailed in a great ship, rowed by 150 lion-hearted men, on the way to the great mines of Punt, far beyond Nubia and south even of Ethiopia. When a powerful storm blew up, the sailor grabbed a piece of wood and flung himself into the sea just as the waves smashed the ship to bits, killing all on board. At daybreak, the sailor realized he had beached on an island of plenty, an Eden where figs, grapes, herbs, berries, fish, and birds abounded.

He satisfied his hunger and comfortably rested in the shade of a broad-leafed thicket. On the third day on the idyllic island, he thanked the gods with a burnt offering. Suddenly, the earth shook beneath him; trees lashed back and forth as if wracked by an approaching gale. A sound like thunder reverberated across the land as a bearded serpent, 45 feet long with golden scales and eyes like lapis lazuli, coiled itself in front of the traveler and demanded to know why he had come to the island. Terrified, the sailor told the snake the story of his shipwreck. The King of Serpents spared him, convinced that the god Amun-Re had had a hand in his salvation. In four months, the golden monster said, a ship from Egypt would come and carry the wanderer home in safety. If he would wait patiently, he would return to his homeland and hold his wife and children in his arms once again.

His island, said the reptile king, held 76 serpents—his brethren and his children—and contained more perfumes than in all the land of Punt. The little kingdom would fade away when the marooned Egyptian left, but surely the gods would reveal it once more to some other wanderer.

After four months that passed quickly on the enchanted isle, the traveler saw an Egyptian ship approach. He hailed it and, with the blessing of the golden serpent, left with gifts of precious perfumes, cassia and other spices, kohl, cypress, incense, and ivory for Pharaoh Amenemhet. As the vessel departed, the magical island faded; where it once lay, moonlight revealed only windswept waves.

Spices and perfumes of the Near East still are a precious commodity.

The ship, the sailor assured the grand vizier, sailed north, and in the second month reached Egypt; then the traveler made his way across the desert from the Red Sea to Thebes. He urged the grand vizier to take him to the pharaoh so he could deliver the treasures of the King of Serpents and beg to be made a commander of a royal ship to sail again into the ocean that washes Punt.

Pharaoh Amenemhet was so delighted with the story of the shipwrecked sailor that he commanded his scribe to write it on a roll of papyrus, so that the tale of such a brave and daring mariner would live forever. The story has been preserved on Papyrus Leningrad 1115, now in the collection of the Hermitage in St. Petersburg, Russia.

THE FIRST RULING *queen of Egypt, Hatshepsut, seized power on the death of her husband Pharaoh Thutmosis II in the 15th century B.C. She sent a trading party on a journey from the Red Sea to the distant land of Punt in search of ebony, gold, and ivory. Reliefs of single-masted galleys rowed by 30 oarsmen carved on the walls of her spectacular mortuary temple in the cliffs west of Thebes recall the expedition.*

Despite Ramses' victory over the Sea Peoples, unrest in Egypt continued to grow. With the death of Ramses III, the country declined further, and about 1166 B.C. it split into two parts, ruled by foreign dynasties in the north and by a series of high priests of Amun in the south. Egypt's wealth waned, its empire in Asia disintegrated, and its prestige elsewhere abroad was lost. By 1070 B.C. Egypt was reduced to its original borders along the Nile Valley.

Kings from Egypt's southern neighbor, Kush, in the Nubian desert of today's Sudan, took power and ruled Egypt from the mid-eighth to the mid-seventh centuries B.C.

In 663 B.C. Assyrian invaders advanced into Egypt, sacking Thebes of its centuries of accumulated treasure. Vassals of the Assyrians ruled Egypt at their discretion, and the country was vastly changed. Foreigners, especially Greeks, were settling along the Nile as traders or mercenaries.

When the Assyrian Empire fell to Babylonia at the end of the sixth century B.C., King Psamtek I reasserted Egyptian power. From the western delta town of Sais, he subdued neighboring princes and reopened the Nile and desert trade routes. Psamtek's son, Necho II, established the first Egyptian navy and began construction of a canal connecting the Nile and the Red Sea.

Egypt continued to compete in the Mediterranean trading arena, but repeated skirmishes with Babylon weakened the country.

In 525 B.C. Persia invaded Egypt; power seesawed back and forth as the Egyptians revolted. In 343 B.C. the Persians attacked again and took control from Nectanabo II, the last independent Egyptian pharaoh. Few Egyptian sources exist that describe everyday life during this period, but the implications are that it was grim.

When Alexander the Great, the 24-year-old king of Macedonia challenged Persia and conquered Egypt in 332 B.C., its inhabitants welcomed him as a liberator. His arrival, however, marked the end of the 3,000-year-old

LAST GREAT KING *of Egypt, Ramses II indulged his passion for building in a colossal temple at Abu Simbel, on Lake Nasser. Four 67-foot statues of the king dwarf his family below him and the god Amun-Re above the entrance. Figures of the gods guard the inside of the shrine.*

kingdom of Egypt; Hellenism grew in power and in influence, eclipsing Egyptian customs and culture. Under Alexander, who also took on the title of Egyptian pharaoh, Greek mores spread through the empire. Egyptian culture survived underneath a blooming overlay of Hellenic customs, art, architecture, city planning, and learning.

One of Alexander's first projects upon annexing Egypt was the construction of a namesake city and seaport near the westernmost mouth of the Nile River. Planned by a Greek architect, Alexandria was enclosed in nine miles of walls, and its rectangular grid of streets was divided into regular blocks. The site had a fine harbor, but the Greeks improved upon it. An artificial breakwater built from the mainland to an offshore limestone islet called Pharos created two sheltered basins accessible under various wind conditions. Lake Mareotis, which bounded the city on its inland side, was connected to the Nile by a canal, allowing Alexandria's great harbor access to southern trade routes. The port opened Egypt to Mediterranean trade more than ever before; glass manufactured in Alexandria, for example, traveled as far as China. A great shipbuilding center established at its docks accommodated hundreds of ships. From its harbor, sea trade with the eastern Mediterranean and the Black Seas flourished.

Ptolemy, a commander in the army of Alexander the Great, became administrator of Egypt upon Alexander's death. When the Macedonian Empire broke apart soon thereafter, Ptolemy declared himself pharaoh in 305 B.C. He and his descendants continued to rule Egypt until its fall to Rome in 30 B.C.

Ptolemy I and his son, Ptolemy II, established sea routes to India via the Red Sea, the Arabian Gulf, and the Indian Ocean. Under the Ptolemies nearly all land belonged to the pharaoh, who grew rich from taxes on imports and exports. Ptolemaic Egypt was a Greek world. Although they were allowed to continue to practice their traditional religion, Egyptians had virtually no representation in civil or military organization. They were separated socially as well as physically from the growing numbers of foreigners, who built their own settlements or established their own quarters in existing cities.

Alexandria became a center of Hellenistic learning and knowledge. To encourage harmony among the many Greek immigrants, Ptolemy I established a new cult based on a god called Sarapis, which was an amalgamation of Greek and Egyptian gods. He founded the Sarapeion, a temple to the new god, and the cult of Sarapis grew popular throughout the Mediterranean. The king founded a library and a museum in Alexandria. Ptolemy II enlarged the collections; the library grew to contain more than a half million parchment and papyrus scrolls that contained the wisdom of the ancient world.

Alexandria's location was ideal for trade, but the coastline was flat and treacherous. The Ptolemies built and dedicated the first known lighthouse on the island of Pharos. Rising more than 330 feet in three levels from a massive stone base, the Pharos Lighthouse was famous as one of the Seven Wonders of the Ancient World. Polished mirrors of iron or brass reflected light from a huge fire in the base to guide ships 30 miles out to sea to the new city.

By 100 B.C. Alexandria had a population of half a million, but the power of the Ptolemies was fading. The native Egyptian population became increasingly hostile toward its foreign overlords, and the economy began to flag. Revolts and lawlessness spread throughout Egypt, advancing its decline. Cleopatra VII, the last of the Ptolemies and infamous queen,

FERTILE FIELDS, *the breadbasket of ancient Egypt, edge the Nile (opposite). The lifeblood of Egypt, the river also served as its main highway. Farmers depended on the annual flooding of the Nile to enrich their land. Navigable for 500 miles from the first cataract north to the Mediterranean, the river carried passengers and commerce throughout the land of the pharaohs.*

chose the bite of an asp to end her life rather than become a Roman ornament. After her and Marc Antony's defeat at Actium in 31 B.C. to the forces of Octavian (later known as the Roman Emperor Augustus), Egypt became a province of the Roman Empire. Alexandria continued to be an important port city during the Roman period.

Alexandria's glories came to life again in the 1990s, when Jean-Yves Empereur, a director of research at France's National Center for Scientific Research and founder of the Center for Alexandrian Studies, was invited by local authorities to begin excavations at various sites threatened by development. Although numerous literary sources mention Alexandria, few descriptions of the ancient city exist. The results of more than a century of archaeological research are scant for a number of reasons. In the 19th century, a time when Alexandria was rapidly expanding after its long decline, archaeologists bypassed the city for Cairo. For many years following, excavations were led by local authorities with limited resources. Ancient strata lay deeply buried, often underwater due to the rising water table in the area. Almost no physical trace of the ancient port exists; no temples remain upright, and its most famous remains—the ancient cemeteries—actually lie outside the city's walls.

One of the few visible remains of the ancient city of Alexandria is a monolith of Aswan granite capped with a Corinthian capital known as Pompey's Pillar. Jutting from a hillside, the 98-foot-high column was a landmark for centuries, leading sailors to the city and tempting daredevil tourists to climb it. Napoleon Bonaparte watched from its base as his troops took Alexandria for France in 1798.

In the 1960s some tombs were excavated in the city center, but urbanization and development hindered research;

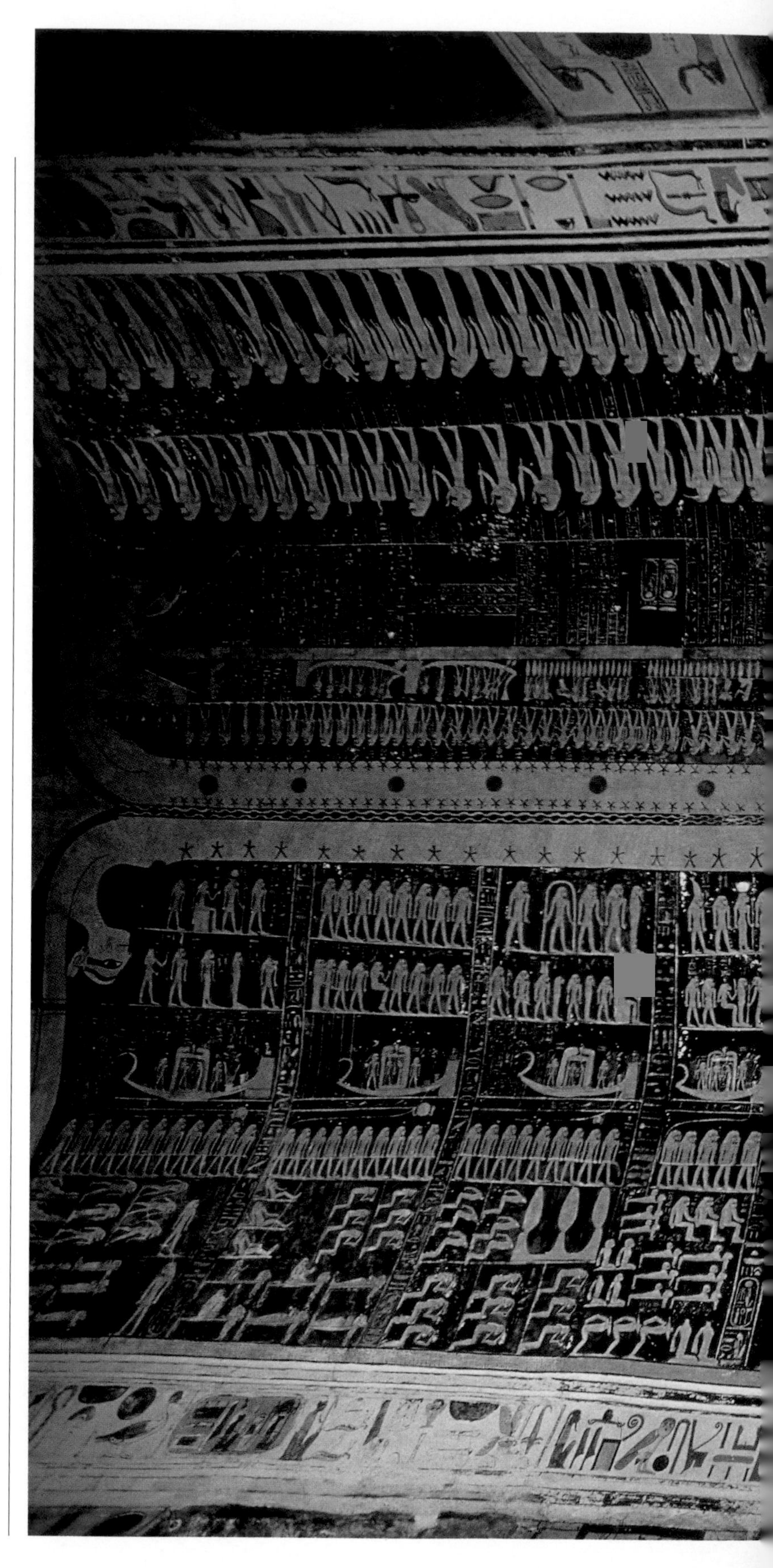

THE SUN GOD *sails through the underworld in a ceiling painting of the tomb of Ramses VI in the Valley of the Kings; boats crossing a river featured prominently in Egyptian mythology as mortuary spells to accompany the dead.*

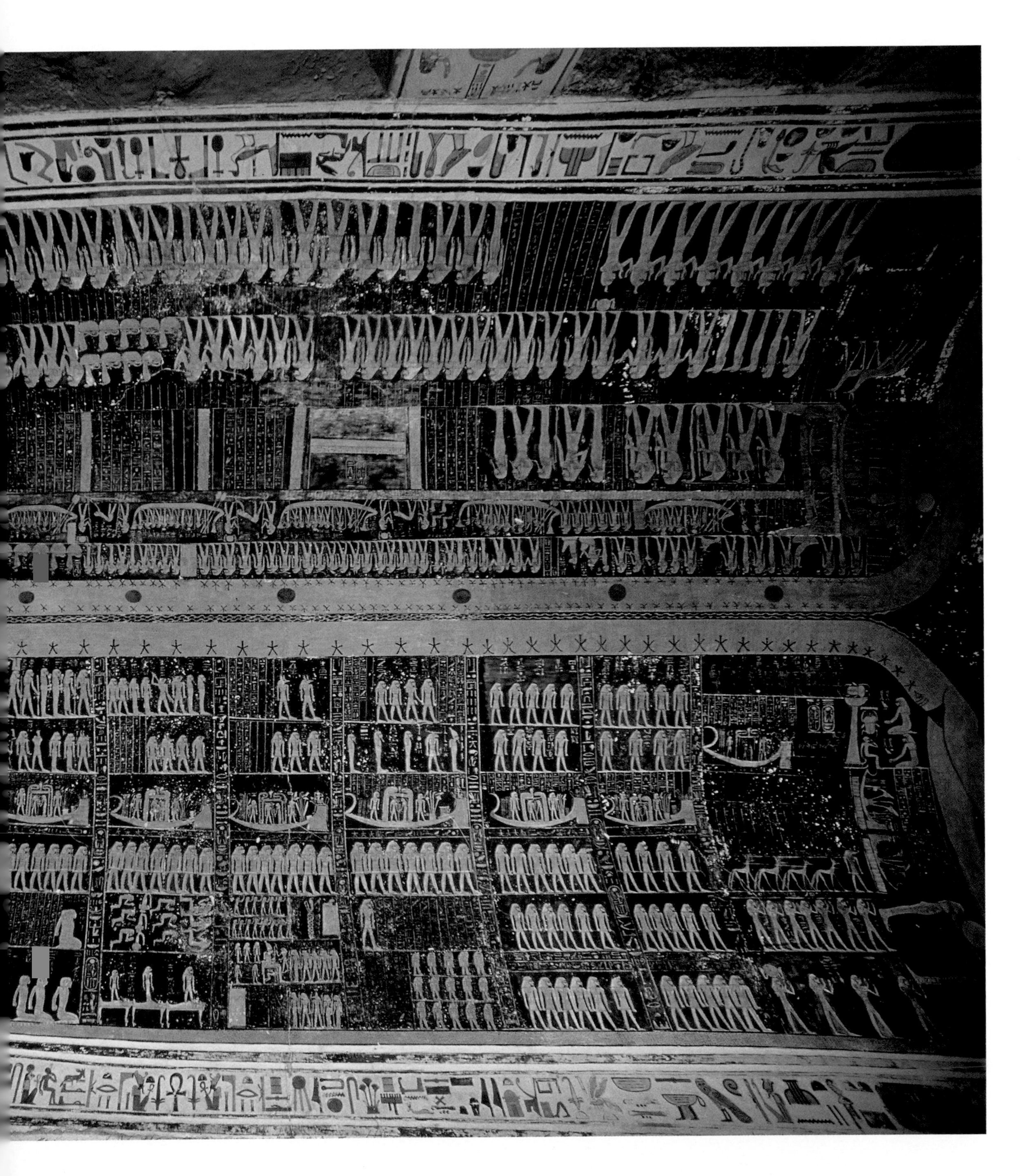

archaeologists could dig only in certain areas for a short time. Nothing remains of Alexandria's world-renowned library, burned by a mob of rampaging Christians during the fourth century A.D.; the Pharos Lighthouse was tumbled by earthquakes by the 14th century A.D.; and the legendary tomb of Alexander the Great has never been found. An archaeological park surrounds the few remains of the Sarapeion; one of the subterranean galleries passes below Pompey's Pillar.

Still, Empereur and his team shed new light on much of the ancient city. They excavated houses of the Alexandrians and their necropolis and researched the system of cisterns and canals that supplied the site with its only source of fresh water. In catacombs at Gabbari, west of the ancient city walls, hundreds of skeletons found in burial niches will enable scientists to study a cross section of Alexandria's Greek population. Near Lake Mariut, ancient Mareotis, which forms Alexandria's southern boundary, excavations revealed wine presses, remnants of vineyards, and pottery workshops for the production of amphorae, proof of a burgeoning wine trade so extensive that a wreck discovered off Marseilles, France, contained dozens of the same amphorae. Off the Islamic fort of Qait Bey, built in the 15th century from ruins of the Pharos Lighthouse, Empereur discovered remains of ships that sank during the Greco-Roman and Byzantine periods. They carried amphorae and pottery that document Alexandria's trade from the fourth century B.C. to the seventh century A.D. Located within some 1,200 feet of the

THE CRESCENT *of the western breakwater girds the eastern harbor of Alexandria, today a large basin anchorage for fishing boats (left above); Fort Qait Bey, built in the 15th century A.D., dangles like a pendant from the curve. Beyond it, archaeologists have recovered submerged remains of the Pharos Lighthouse; an old engraving depicts it as it may have looked then (right above).*

shoreline, the vessels probably foundered on the almost invisible reefs that partly obstructed the harbor.

In 1996 Empereur sighted perhaps his most exciting find. Sunken near the base of Fort Qait Bey some three thousand huge stone architectural blocks carpeted more than five acres of the seabed. Empereur's team also unearthed hundreds of huge columns of pink Aswan granite and fragments of obelisks and sphinxes dating to the time of the pharaohs. Huge male statues hewn from the same pink granite, effigies of the Ptolemies, probably stood at the foot of the Pharos. Door jambs and lintels of the same material—blocks up to 38 feet high and weighing up to 75 tons—were identified as part of the Pharos Lighthouse, flung wide in a violent earthquake.

Empereur's efforts brought to light the sunken remains of one of the world's longest-standing functional monuments. For nearly 16 centuries the Pharos Lighthouse served the sailors and traders who fostered Egyptian trade. The civilization itself lasted far longer—more than 3,000 years—and its influence is felt to this day. ■

SUNKEN TREASURES OF ALEXANDRIA

Little remains standing of ancient Alexandria, Egypt, once one of the greatest ports in the world, center of learning, and a rival of Athens and Rome. In the 1990s French archaeologist Jean-Yves Empereur began excavations on land and sea that recalled the city's glorious past.

Empereur and his team unearthed Alexandrian houses and graves, added knowledge of the engineering that provided fresh water to the city, and uncovered relics of its Mediterranean wine trade.

Off the 15th-century Islamic fort of Qait Bey, constructed from stone used to build the ancient Pharos Lighthouse, Empereur found some three thousand architectural blocks scattered over five acres of the seafloor. Lintels and door jambs of pink Aswan granite, blocks weighing up to 70 tons, proved to be from the Pharos, the 330-foot ligthhouse that served Alexandria for nearly 1,500 years and was one of the Seven Wonders of the Ancient World. They had been loosed by one of the many earthquakes that finally brought the lighthouse down in the 14th century A.D.

Divers began recovery of the relics in 1995. At left, a diver, accompanied by friendly fish, sketches a gray granite sphinx. Hieroglyphs on the base of the statue, the only one found with its head intact, indicate that it dates to the reign of Ramses II.

Empereur and his team recovered some 25 sphinxes composed variously of gray granite, pink granite, and calcite. Several columns carved to represent papyrus also carried the cartouche—the hieroglyphic name—of Ramses II, dating to nine centuries before the founding of Alexandria. Like the sphinxes and many other of the underwater monuments, the columns were brought to Alexandria from an

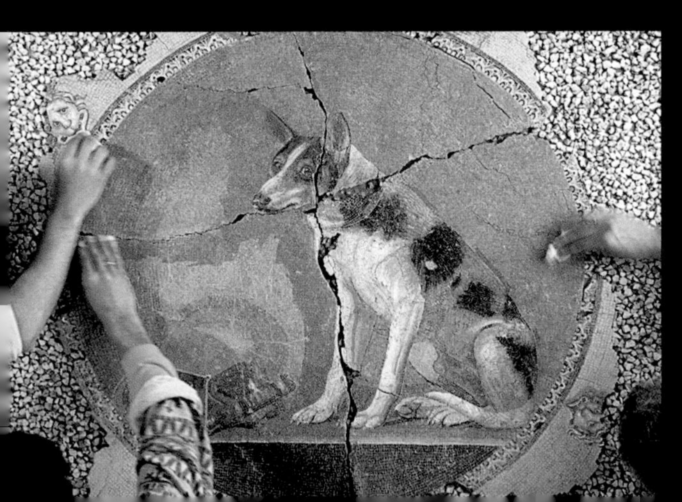

abandoned sanctuary at the ancient capital Heliopolis.

A 17-ton Aswan pink granite fragment hoisted from the seabed in 1995 dwarfs the modern skyline of Alexandria as well as the recovery boat (right). The stone proved to be the torso of a statue of one of the Ptolemies represented as a pharaoh that probably stood at the foot of the Pharos Lighthouse.

Hieroglyphs depict the pharaoh bringing offerings to the gods at Heliopolis on the upper part of a calcite obelisk of Seti I, father of Ramses II (above left). It was one of hundreds of obelisk fragments and columns recovered by Empereur and his team.

Finds on land were no less impressive. Excavations of the Egyptian Archaeological Services on the site of the future Biblioteca Alexandrina bared a mosaic of a spotted dog beside an overturned pitcher (below left).

The animal's red collar indicates that he was a pet; the cracked mosaic is composed of thousands of tiny tesserae only an inch wide and dates to the second century B.C.

CHAPTER FOUR

THE MINOANS

3000 B.C. — 1380 B.C.

Slopes of Santorini, largest remnant of ancient Thera, plunge to the flooded crater of its volcano; beyond lies Néa Kaiméni.

MINOAN MYSTERIES

BY ROBERT D. BALLARD

FROM ABOUT 3000 B.C. THE MINOANS OF CRETE RULED the Aegean for several centuries. They were history's first maritime power, master sailors whose powerful navy freed the seas from pirates. The Minoans traded throughout the Aegean, as far east as Asia Minor, and sailed south to Egypt. But underwater archaeologists have discovered no Minoan ships. Whether they exist and where they lie are among the many enigmas associated with the Minoan civilization. Perhaps the Minoan ships all sank in shallow water and have been destroyed by bacteria. Even then, however, their cargo should still litter the ocean floor. Yet little has been found.

Many experts tie the demise of the Minoans to the gigantic eruption on the island of Thera, north of Crete, in about the middle of the second millennium B.C. Maybe tidal waves, earthquakes, and fire sank and destroyed the Minoan fleet. The question is whether the explosion could have been powerful enough to have an impact on Crete, some 70 miles away.

The great eruption of Krakatoa, Indonesia, in 1883 is well documented. The blast fractured the masonry of buildings a hundred miles distant; ash and dust fell for more than three days on ships 1,600 miles from the site; seismic sea waves, called tsunamis, reached 130 feet in height on nearby coasts, swamping hundreds of villages and killing tens of thousands of people. Scientists have concluded that the eruptions of Krakatoa and Thera were alike in many ways—and that the volcanic explosion of Thera was considerably larger than that of Krakatoa.

Tsunamis reach speeds of several miles an hour. On the open ocean, they may be a hundred miles wide but only a few feet high, barely disturbing ships that sail over them. As they reach shallows, however, the waves crest. An immense wave of water strikes the shore without warning and floods the area within minutes. When a tsunami recedes, it sucks everything out to sea with it. Minoan ships in harbor would have been shattered, then scattered like matchsticks along with their cargo. Seismic sea waves associated with the eruption on Thera could have erased evidence of Minoan culture. At the same time, however, the explosion preserved elements of the civilization it might have destroyed.

At Akrotiri, on the island remnant of Thera now called Santorini, Greek Inspector General of Antiquities Spyridon Marinatos spent years excavating a Minoan seaport, beginning in 1967. Fine dust and pumice 23 feet thick, deposited when Thera erupted, had acted like heavy packing material, preserving the ancient town for centuries. Much of the volcanic ash on Crete, though, has eroded away. Oceanographic vessels have been extracting core samples from the sea bottom in the eastern

Mediterranean since 1974. Ash is consistently thicker in the samples collected nearest Thera.

The island was only one of the Minoans' far-flung colonies. Minoan mariners regularly sailed to Egypt trading wine, olive oil, and fabrics for papyrus, gold, stoneware, and gemstones. Minoan motifs appear in Egyptian palaces and royal tombs in Thebes; silver bowls of Minoan manufacture were part of a pharaoh's funeral goods as early as 2000 B.C. We do not know what the Minoans called themselves; Egyptians called them the "Keftiu," or the "Isles in the Midst of the Sea."

It was the Egyptians who—according to Plato, the originator of the legend of Atlantis—first related the story of a rich and sophisticated island continent that was suddenly destroyed in a natural disaster. In about 350 B.C. Plato wrote: "Afterwards there occurred violent earthquakes and floods; and in a single day and night of misfortune the island of Atlantis...disappeared in the depths of the sea."

Maybe Plato's Egyptians were referring to the island of the Keftiu; perhaps Thera was their Atlantis. And perhaps underwater archaeologists will find Minoan wrecks off the coast of Egypt, where so many vessels from Crete sailed for more than six centuries. ■

"

THE DOCKS [OF ATLANTIS] WERE FULL
OF TRIREMES AND NAVAL STORES,
AND ALL THINGS WERE QUITE READY
FOR USE.... THE ENTIRE AREA WAS
DENSELY CROWDED WITH
HABITATIONS; AND THE LARGEST
OF THE HARBORS WERE FULL
OF VESSELS AND MERCHANTS
COMING FROM ALL PARTS, WHO
FROM THEIR NUMBERS,
KEPT UP A MULTITUDINOUS SOUND
OF HUMAN VOICES
AND DIN AND CLATTER OF ALL
SORTS NIGHT AND DAY.

"

PLATO, "CRITIAS"

COMPU-CENTRAL

Testing equipment for future missions in the command/control center aboard the support ship Knorr *during an expedition in 2003, Ballard studies one of six huge plasma displays in a large room (left), from which scientists relay specific directions to technicians in a second room. The command center—two shipping containers welded together and attached on deck—also contains 16 computer displays, many small monitors, and work stations for 12, and allows the team supervisory control previously unattainable. In the science lab on the ship, engineers wrote the original software code that makes each individual element of the technology work as a system (above).*

EYES WIDE, *an octopus wriggles across a jar from Knossos, Crete. Minoan culture covered three periods based on its pottery. Underwater motifs—the Marine style—date this piece to the 15th century B.C., the height of the civilization.*

THE MARINERS OF CRETE

NO ANCIENT CIVILIZATION IS AS MARKED WITH mystery and magnificence as that of the Minoans of Crete, whose culture developed there from about 3000 B.C. and thrived until its downfall around 1450 B.C. On their island, which separates the Aegean Sea from the Mediterranean, they created one of the great Bronze Age cultures and became the first true maritime power in history. Like other Aegean peoples, the Minoans took to the sea originally to obtain the copper and tin they needed to make bronze. They sailed the Aegean and eastern Mediterranean, trading timber, pottery, gold, ivory, and gemstones. The wealth and influence accrued from trade in the Minoans' maritime empire formed the basis of a sophisticated, busy, and creative society that cherished nature, beauty, and pleasure. Scenes from Minoan art suggest that they were a happy people, preoccupied with sports, religious rituals, dancing, and the beauty of their surroundings. Yet few civilizations are so completely lost in time.

Archaeological excavations have yielded examples of their elegant pottery and jewelry. Lavishly decorated palaces and village houses with indoor plumbing provide ample evidence of their brilliant society, but the Minoans seem to have had no sense of history. Little written record of them exists, and none includes names or dates. Minoan script—termed Linear A—remains largely undecipherable. To this day archaeologists disagree about what disaster wrought their final downfall.

The precursors of the Minoans probably migrated to Crete about 7000 B.C. from Asia Minor. Around 3000 B.C. farmers, herders, and artisans lived in scattered villages in the craggy interior of Crete. They built shrines on mountaintops and buried their dead in communal graves. Arable plains between the island's mountains yielded grain, timber, and olive oil.

At Myrtos, on the southeastern coast of Crete, numerous rooms in a large stone and mud-brick building appear to be a grouping of households, enlarged to accommodate the families through time. The craftsmen of Myrtos threw pots—the Minoans were the first people in the Aegean to use a hand-turned pottery wheel, probably beginning about 2500 B.C.—and they wove and dyed cloth. Farmers cultivated olive trees, vineyards, and fields of barley; they kept sheep, goats, cattle, and pigs. Inhabitants worshiped at a small shrine, the earliest known in Crete, in the southwest corner of the settlement.

Southeast of the Greek mainland, Crete commands the entrance to the Aegean, and Minoan traders made that sea their private lake. The island's natural harbors and its position as a hub of the Mediterranean made it a crossroads of trade routes. The largest concentration of settlements was in the central and eastern part of the island, with villages both in the south, particularly on the Messara plain, as well as in the north.

The Minoans achieved power and wealth because of their command of the seas, and their civilization owed a large debt to those of Egypt and the Near East. The Cretans developed extensive commercial and diplomatic relations with Egypt and the cultures of the Syro-Canaanite coast (modern-day Israel, Lebanon, and Syria). Traders and craftsmen learned new techniques from their commercial partners and then developed them further. Local chiefs of minor settlements gained power and

DOLPHINS *escort ornate galleys in a close-up of a tribute to Minoan sea power uncovered at Akrotiri, on the island of Santorini. The 16-inch-high fresco covered three walls of a house, stretching 20 feet.*

emerged as rulers. After about 2000 B.C. the population of Crete concentrated around several power centers on the island—Knossos, Mallia, Gournia, Zakros, and Phaistos. These settlements began to emerge as palace complexes that grew in power, size, and opulence over the next 300 years. Crete became a land of small city-kingdoms. The Minoans invented their writing system around 1800 B.C., but seem to have used it only for record-keeping.

In the *Odyssey* Homer extols "a land called Crete in the midst of the wine-dark sea, a fair land and rich, begirt with water, and therein are many men innumerable, and 90 cities." The existence of that culture came back to light only at the turn of the 20th century, when wealthy Englishman Arthur Evans traveled to Crete to investigate gems and seals he had bought in Greek bazaars. Inspired by these seals, as well as by mythological references to a Cretan culture ruled by a king named Minos, Evans bought a six-acre site at Knossos, on the northern coast of the island, and began full-scale excavations in 1900.

Within a short time he and his team began to unearth the remains of a huge and luxurious palace. Terraced into the hillside, the multistoried structure surrounded an open courtyard and sat above floors of underground chambers and passages. Thousands of artifacts discovered at the site testified to a life of luxury in a graceful setting that predated classical Greek culture by more than a thousand years.

One of the rare names traceable to Crete in the Greek myths Evans read about was that of Minos, a legendary king of the island who kept a monster, half-bull half-man, called the Minotaur. Evans assigned the name "Minoan" to the civilization, one whose existence was known only from myth. Much archaeological activity followed, and renewed exploration revealed additional palace complexes and comfortable villas scattered over the central and eastern parts of the island.

In more than 20 years of excavations at Knossos, Evans unearthed a sumptuous palace of interconnected rooms

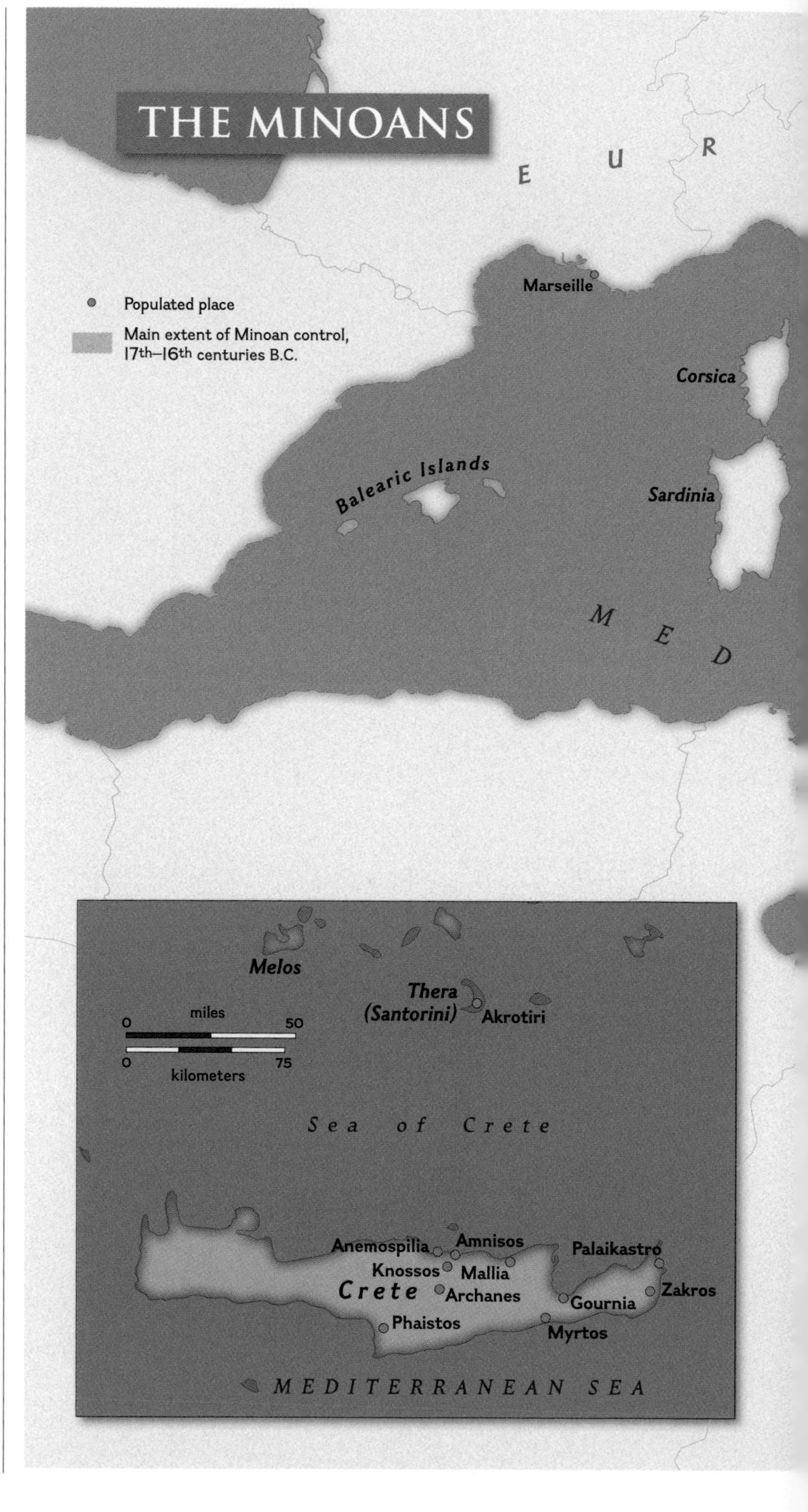

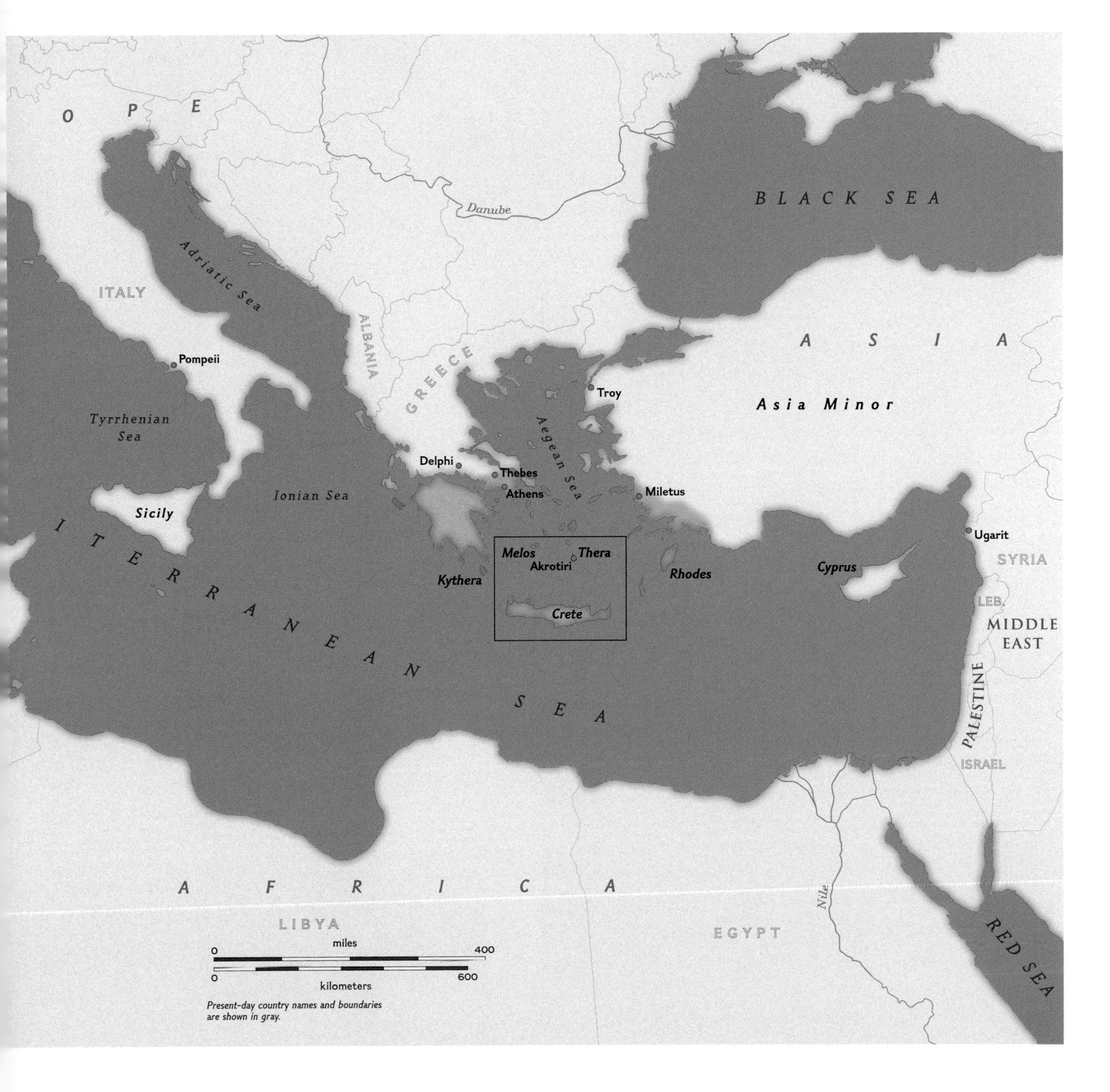

O P E
BLACK SEA
Danube
Adriatic Sea
ITALY
ALBANIA
GREECE
Pompeii
Troy
A S I A
Asia Minor
Tyrrhenian Sea
Aegean Sea
Delphi
Thebes
Athens
Miletus
Ionian Sea
Sicily
Ugarit
Melos
Thera
Akrotiri
SYRIA
Kythera
Rhodes
Cyprus
LEB.
MIDDLE EAST
Crete
I T E R R A N E A N
S E A
PALESTINE
ISRAEL
A F R I C A
LIBYA
Nile
EGYPT
RED SEA
miles
0
400
0
600
kilometers
Present-day country names and boundaries are shown in gray.

decorated with detailed frescoes and furnished with finely engineered indoor plumbing and luxurious ceramic bathtubs. An elaborate system of drains served flushable toilets.

As he dug, Evans reconstructed parts of the palace. Although his mixing of its ancient stone with concrete and steel drew criticism, his reconstruction efforts made Knossos one of the few Minoan sites where easy visualization of the ancient civilization is possible. The interior space, much of it windowless, is a maze of royal apartments, interconnected suites, and large storerooms, all arranged around a central court. From column bases remaining in Knossos, Evans deduced a unique feature of Minoan architecture. Round columns, unlike those built by the Greeks or the Egyptians, grew thicker as they rose from the ground. The frescoes that Evans uncovered on palace walls offered a first look at the dark and elegant Minoans themselves. Men were always painted in red, and women—painted with white skin—were slender and graceful, with narrow waists and carefully arranged coiffures. A breechcloth or short wrapped skirt was standard attire for bare-chested Minoan men; women wore long flounced skirts and short open bodices. The women, as the frescoes imply, were much admired. They mingled freely with Minoan men, enjoying a remarkably emancipated position in their society. Even many years later, in progressive Athens, women would not enjoy such freedom.

Based on changes in pottery styles, Evans divided Minoan culture into three major periods. More recent research has honed his dating, and now the periods are usually referred to in relation to the palace-based structure of the Minoan society. During the Prepalatial period,

BAKED CLAY *amphorae and other pottery vessels helped archaeologist Spyridon Marinatos date his discovery on the island of Thera at Akrotiri when he began excavations there in 1967. Jars like these—photographed as they lay when excavated—lined the streets and probably stored olive oil, grain, and wine.*

PERHAPS *the first "portable" artwork, a three-foot high stucco panel from Akrotiri portrays a young fisherman bearing a bountiful catch of dolphin fish (opposite). One of the best-preserved frescoes, it radiates the naturalism and spontaneity characteristic of Minoan art. Another fresco discovered intact beneath the volcanic debris depicts a river landscape (above).*

roughly during the third millennium B.C., the culture established itself on Crete.

Metalworking became widespread, and craftsmen began to specialize in particular metals or precious stones. Inhabitants of the island started to cultivate vines and olives, to breed sheep for wool, and to build roads, bridges, and aqueducts.

During the Protopalatial (or Old Palace) period, Minoan cultural and social organization developed further, and palaces like Knossos were built on the island. The population increased; agriculture, manufacturing, and trade reached new heights. Merchants took to the sea, consolidating trade links with Anatolia and establishing new ones with the eastern Mediterranean and Egypt. The Minoans, termed "Keftiu" or the "Isles in the Midst of the Sea" by the Egyptians, probably traded lumber, olive oil, pottery, and cloth for papyrus, stonework, and gold. At the temple of Tod, in the vicinity of Thebes, silver vessels of Minoan style were found in chests, two of which were inscribed with the name of the 12th-dynasty Pharaoh Amenemhet II. A list of Aegean place names was discovered in a courtyard of the mortuary temple of the Pharaoh Amenhotep III at Kom el Heitan near Thebes. Tableaus in tombs of nobles show Minoans presenting their wares to

the Egyptians during the reigns of Queen Hatshepsut and Pharaoh Thutmose III. The rise of Minoan writing—called Linear A—attests to the diversity and complexity of Minoan society.

Based on the lack of fortifications at palaces and towns, Evans postulated that the Minoan society was a confident, powerful, and peaceful one, and he looked for a natural cause for the disruption of their civilization. His theory—that a series of earthquakes followed by catastrophic fires leveled the palaces—was accepted for decades. Little evidence remains of the original structures such as the one at Knossos, but the disaster that tumbled the palaces did not halt growth on Crete.

COBBLES *delineate the main street in Gournia (opposite), a village on the north coast of Crete. The six-inch-wide Phaistos Disk (above), found about 1900, holds the earliest example of writing but remains indecipherable.*

The palaces were rapidly rebuilt in even more splendid form, and the great age of Minoan Crete, termed Neopalatial period, rose from the rubble. Minoan culture began to peak with the reconstruction of the palaces and remained in full bloom until about 1450 B.C.

In the fifth century B.C. Greek historian Thucydides wrote in his books on the Peloponnesian Wars that King Minos of Crete was the first ruler to build up a navy and to establish a maritime supremacy. Although these claims have been heavily disputed, archaeologists have documented a strong Minoan influence in the southern Aegean, perhaps motivated by the Minoans' search for metals. According to Thucydides' writings, King Minos also freed the Aegean from pirates. The lack of palace fortifications on Crete has been explained as a result of Minoan naval superiority. Clay models and drawings of ships on seal stones reflect these Minoan interests. The typical vessel was small, with oars and a center mast that carried a square sail; the ships had a crescent-shaped hull, a high prow, and a low stern. The prevailing northeasterly winds in the Aegean dictated that their trade route was circular. They struck out across the Mediterranean, sailing blue water to Egypt. Then, by employing land and sea breezes they worked their way back to Crete by sailing up the Syro-Canaanite coast and following the southern coast of Asia Minor (modern Turkey). Asia Minor produced tin, Cyprus offered copper, and Egypt exported luxury goods such as gold, ivory, and glazed clay ornaments.

Wealth from trade was lavished on the sumptuous Minoan palaces and the paved roads that linked them. Mazelike Knossos was the most important of the Minoan palaces. The sprawling palace complex covered five acres. On the coast to the north of Knossos lay Amnisos, the palace harbor.

Like other Minoan palaces, Knossos was much more than a royal residence. It was also an administrative hub, manufacturing center, storage depot, and the heart of the area's religious life. Minoan ships, hailing from Amnisos and other ports, sailed with items of Cretan manufacture, rightly renowned throughout their sphere of contact, and returned with valuable trade items. Like all Minoan palace centers, Knossos drew tribute from farmers and herdsmen.

Produce paid as taxes was funneled into the palace, stored, and redistributed, mostly as trade items. The centralization of power, bureaucracy, and wealth on Crete gave rise to palace economies that were pivotal to Minoan society.

Who ran the palaces and supervised distributions remains a mystery. As in all ancient kingdoms, the rulers in Crete were the embodiment of the state, representatives to the gods of the land and the people. Most authorities suspect that the ruler of each palace complex was a priest-king, but no archaeological evidence as yet supports the theory. It does appear, however, that the society was highly stratified, and that the redistributive system was very efficient. Each palace controlled the production of manufactured goods and food, ensuring sufficient wealth for the production of luxury goods. At Knossos and other palace complexes, large areas devoted to storage and workshops as well as the impressive network of roadways indicate the importance of commerce to Minoan civilization. Surplus crops of grain and oil as well as wool, flax, and hides were stored in the palace complexes along with raw materials imported from neighboring countries that artisans would craft into material goods.

The essential plan of a Minoan palace complex was standard. A maze of public and private rooms, bathrooms, kitchens, storerooms, and cult rooms were arranged in uneven blocks surrounding a central courtyard. Excavations of their ruins testify to the Minoans' remarkable knowledge of engineering. Faced with the great swing in seasonal temperatures on Crete, with stifling, hot summers and raw, cold winters, Minoan architects used open courtyards, light wells, and connecting doorways, which afforded cross-drafts to make life in the palaces tolerable in hot weather, but could be closed off during inclement weather. In addition to their complex palaces and paved roads, the Minoans constructed viaducts, aqueducts, drainage systems, and irrigation channels across Crete. The villas were airy and spacious as well, with plentiful water supplies and brightly painted walls. The elite class enjoyed enormous luxury and finely crafted leisure items such as inlaid checkerboards. Beyond the rich lifestyles of the royal family and nobility existed several social classes. Well-off townspeople probably represented a small segment of the free population. Their houses and villas were usually built in small towns with narrow, winding streets. Almost no archaeological trace remains of the thousands of farmers and craftspeople who served the upper classes. The village of Gournia is a lone example in helping to understand Minoan life outside the palaces. Two-storied houses lined narrow streets. On the ground floor, magazines and workshops held the tools of everyday life for farming, fishing, potterymaking, and weaving. The living quarters were above the stores.

Minoan artifacts, although rare, are found throughout the eastern Mediterranean. Settlements that are either Minoan, or highly influenced by the Minoan culture, are found on Aegean islands as well as on the western coast of Asia Minor and Egypt.

In their pottery as in their jewelry, metal, and ivory work, the craftsmen of Crete were skilled at conveying a sense of life and movement. A feeling of spontaneity pervades all Minoan art. Martial themes are rare; rather than choosing historical or political themes, Minoan artists emphasized the environment and reflected a love of nature in their works. Inlaid daggers and engraved axes found in tombs and sacrificial sites reveal that metalwork during Minoan times reached new heights. Artisans also excelled in producing exquisite carved gem seals, gold and ivory miniatures, and frescoes. Some of Evans's most exciting discoveries at Knossos were wall paintings which brought Minoan civilization to life. Like the rest of Minoan art, the frescoes reflect a vivid picture of a talented, luxury-loving, and highly sophisticated society. Fragments of Minoan-style wall paintings in Israel and Egypt suggest that artists from Crete traveled far and wide, displaying their skill.

THE LEGEND OF THESEUS AND THE MINOTAUR

When the god Poseidon sent a sacrificial bull to the island of Crete, the wife of the legendary and powerful King, Minos, fell in love with the bull. The result of their union was a dreaded monster—the Minotaur—with the body of a man and the head of a bull. To hide this disturbing creature, King Minos ordered the architect Daedalus to build a labyrinth for the monstrous offspring.

The Minotaur lived solely on human sacrifices: Seven youths and seven maidens sent to Crete annually by the city of Athens as a tribute to Minos after his son had been killed in Athens.

One year Theseus, the son of Athenian King Aegeus, having returned from a stay in the country, found the city preparing to send its requisite sacrifices to Crete. The brave youth offered to take a place among them and attempt to kill the insatiable Minotaur. Theseus sailed for Crete on a ship rigged with a black sail. "If I succeed," he promised, "I will return under a white sail, advertising my success."

King Minos and his lovely daughter, Ariadne, met Theseus and the other doomed youths when their ship landed in Crete. Ariadne, much taken with the bravery of Theseus, offered to help him in his self-appointed task, if he would take her away from Crete and marry her. Theseus agreed. Ariadne described to him the sloping corridors, chambers, and stairways of the labyrinth—a huge maze from which no one had ever escaped.

That night Ariadne sneaked to the prison room of Theseus and, with her own key to the palace, released him. She guided him to the labyrinth and pointed to a small door. Then Ariadne gave Theseus a torch, a sword with which to slay the Minotaur, and a ball of wool with a weight attached to it. When she was a little girl, she told Theseus, the builder of the labyrinth showed her how to find her way out of the maze. She advised Theseus to tie the end of the yarn to the door lintel and allow the ball to roll down the sloping floors of the labyrinth until it reached the center—the lowest point in the building. There the Minotaur would be found—probably asleep at this time of night.

Bull's head and male body on a torso of the Minotaur remain; time has clipped its horns.

Prince Theseus of Athens slays the bull monster in an 18th-century print inspired by a Greek vase painting.

Theseus proceeded exactly as Ariadne had directed and followed the gently rolling skein of wool down the passages. The air grew thicker and warmer as he approached the center of the maze, and an animal stench announced the monster. On a golden bed, half covered with straw and surrounded by gnawed bones, Theseus found the Minotaur fast asleep and snoring loudly. The monster awoke as Theseus approached and rushed at him, bellowing with rage, but the hero attacked with his sword, slashing and slicing until the half-man, half-bull fell dead.

Wounded and weary, Theseus carefully followed the woolen thread back to the entrance of the maze, where Ariadne waited. Together they quickly freed the other captives and hoisted sail in the darkness. In two days they reached the island of Naxos. Fearing for her safety in Athens, Theseus abandoned Ariadne as she lay sleeping. She awoke alone, weeping, but before long a merry procession led by Dionysus, god of feasting, wound its way past. By morning, Dionysus had won Ariadne over; she would marry him and forget her Athenian love.

Theseus, however, was so preoccupied with the fate of Ariadne that he forgot to hoist his white sail as he sailed home. King Aegeus, spying the mournful sail in the distance and fearing his son dead, stumbled, weeping, from a cliff and fell to his death in the sea. Upon landing and hearing of the tragedy, Theseus was overcome with grief. He directed the building of a huge monument to honor his father, and the blue waters surrounding the city are named the Aegean Sea in his memory.

Minoan art also reflects the society's preoccupation with religion. Murals, gem seals, and jewelry are among the many items that depict religious observances. The Minoans did not build public temples or other large worship centers. In groups, they performed religious rites outside on mountain peaks, in rock shelters, and in caves, sacrificing animals and performing dances. Indoors, they erected small shrines in the palace complexes and in the elaborate villas of the Minoan nobility. Two themes—the bull and the double-headed ax—dominated Minoan religious symbolism.

The bull, a symbol of strength and male potency, is a recurring motif on Minoan vases, in frescoes, and other art. Like many elements of Minoan civilization, the cult of the bull probably originated in the East. Numerous depictions hint that bull leaping was practiced by the Minoans. Evans revealed a fresco at Knossos that shows three figures involved in bull jumping. Girls as well as boys performed in the sport. One participant ran headlong at a bull, grabbed its horns, executed a handspring over its head to land on the animal's back, then quickly jumped down into the arms of a waiting assistant. Such feats may have inspired the myth of the Minotaur. Why the bull games took place is, like much of Minoan culture, shrouded in mystery; they may have had ritual significance or may have been simply a popular sport. Where the games took place is also debatable. Some experts believe that the events occurred in the central palace courts. Others point out that those were much too small to host such an activity.

The double-headed ax, as pervasive as the bull in Minoan ritual, was the symbol of the dynasty of King Minos. The Greek word for the symbol of the double ax is believed to be the root for the word "labyrinth." Knossos and other palace-centers, with their mazelike conglomerations of halls, chambers, and courts, resembled labyrinths, and the word is tied to the ancient legend of Theseus and the Minotaur. The double-headed ax, like the cross of Christianity and the crescent of Islam, symbolized Minoan religion. At Knossos, where the design appears repeatedly throughout the complex, Evans discovered a three-level shrine centered around the symbol.

Based on excavations of figures at Knossos and other sites, archaeologists originally believed that Minoan religion was based on the worship of one or more mother goddesses who were associated with animals. More recent research and excavations have implied a greater diversity in Minoan religious practices, with individual communities recognizing different and very local gods. Blood sacrifices were a standard element of festivals and funerals; they centered normally on animals, usually bulls. Near the present-day village of Arkhanes, four miles south of Knossos, Greek archaeologists discovered in 1979 a dark side to Minoan religion—possible evidence of human sacrifice, although this interpretation has been questioned.

The tholos—a low, round stone tomb—is one of the most remarkable features of Minoan culture. More than 80 of the graves have been excavated, most near the palace-center of Phaistos. The tombs, which date from as early as 2800 B.C., long before the rise of palace-states, seem to have been built next to settlements. They were used communally, some continuously used for more than a thousand years. They gradually filled with skeletons and grave goods as they were used over and over again. Periodically the tombs were cleaned and reused. Although all have been plundered of their most valuable goods, a few baked clay models unearthed portray such Minoan practices as dancing and ritual libations.

About the middle of the second millennium B.C., the neighboring island of Thera, also known as Santorini, was convulsed by massive volcanic eruptions that turned the island into a huge cauldron. The event collapsed the central part of the island into the sea and left only a portion of the rim of the volcanic crater behind. Since then the resulting caldera, about 2,600 feet deep and five miles in

WHEN SIR ARTHUR *Evans began excavations on the northern coast of Crete in 1900, he discovered a previously unknown Bronze Age culture. He dubbed them the Minoans, after the legendary King Minos. Knossos, the largest and most powerful of the Minoan palace complexes, once housed more than 40,000 people (opposite). Like other Minoan palaces, Knossos served as an administrative hub as well as a royal residence with more than 1,500 rooms. A painted earthenware casket from the excavations (left) dates to the 15th century B.C.*

diameter, has bubbled up from numerous small eruptions on the seafloor, producing a small new island in its center.

The eruption took place in what archaeologists call the end of the Late Minoan IA period, and there are two main schools of thought regarding the actual date of the eruption. Some archaeologists, basing their views on the archaeometric evidence, prefer a date of 1628 B.C.; whereas others, basing the time primarily on the evolution of pottery styles, date the eruption to circa 1500 B.C. These eruptions spewed an enormous amount of volcanic ash high up into the atmosphere, which may have affected global weather patterns.

Some archaeologists contend that Minoan civilization may have been severely challenged by the eruption and resultant tsunamis—huge sea waves—which may have caused the Minoans' demise. Others have suggested that the Minoans were done in by internal dissent, which left them open to invasion. What is clear is that Crete was eventually taken over by Greek-speaking Mycenaeans from mainland Greece.

Beginning with Arthur Evans, archaeologists have sought an external cause, such as a natural disaster, to explain the abrupt collapse of Minoan civilization. In 1939 Greek archaeologist Spyridon Marinatos hypothesized that the fall of the Minoans was connected to huge volcanic eruptions on Thera, accompanied by earthquakes, seismic sea waves, and ash fall. Tsunamis could have damaged Minoan settlements on Crete, Marinatos suggested, and dense ash fall could have blanketed crops and fields.

Although the evidence indicates that the Minoans never ruled mainland Greece, they had a particularly strong influence on the neighboring culture of the Cycladic Islands north of Crete. Marinatos, convinced that the fall of the Minoan civilization on Crete had been caused by the eruption on Thera, began excavations at the site of Akrotiri, on the southern end of that island. At the

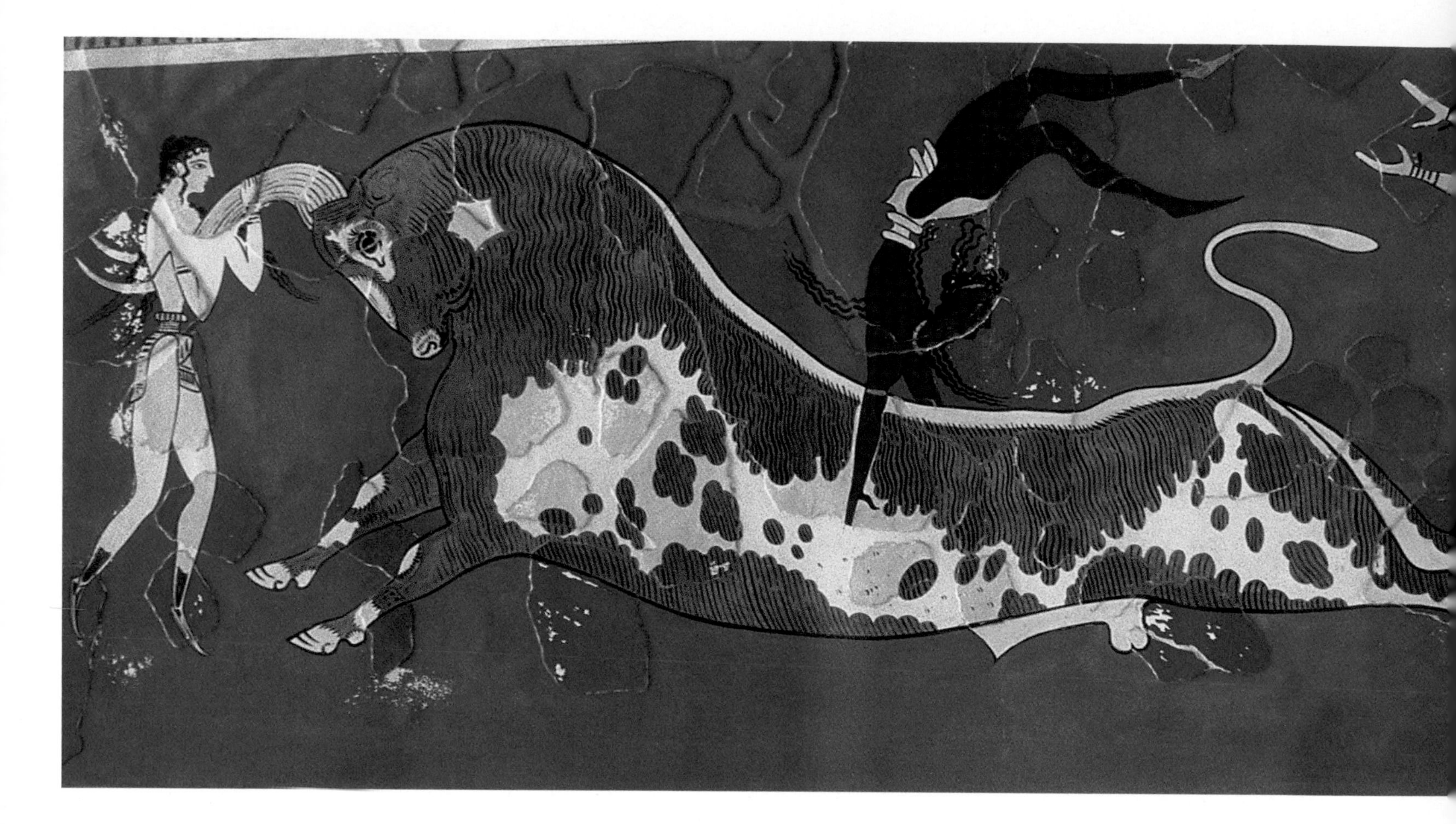

site, a once-thriving port, Marinatos discovered that tephra —fine dust and pumice—23 feet thick covered the land, and layers of ash and volcanic glass had preserved evidence of a highly Minoan-influenced Cycladic culture. Unlike Minoan sites on Crete, Akrotiri on Thera survived under the ash exactly as it was the day disaster struck. The city had been cleaned out in haste and abandoned. Everything of value had been removed. No human treasures or distorted skeletons remained as in the ruins of Pompeii: The inhabitants had been warned, perhaps by earthquakes that presaged the eruption, and salvaged everything of value.

Therans, Marinatos discovered, enjoyed a diet rich in seaweed, liked roasted snails, and ate several different kinds of beans as well as barley and millet. Sesame seeds were used as condiment. Grapes grown on Thera yielded wines known throughout the Mediterranean. Marinatos found remains of terra-cotta plumbing and town houses. A bed he discovered in one of the houses indicates that, like most ancient Mediterranean peoples, Minoans were short—no more than about 5 feet 4 inches tall. Fertility idols of marble and clay reflected the importance of religion to the inhabitants.

Minoan-style frescoes at Akrotiri, which Marinatos unearthed, proved to be his most important finds. More varied and complete than anything discovered on Crete itself, the art revealed details of Minoanized life. Shepherds wear fur caps; women carry jars on their head; boys are portrayed boxing, which may have been a sport related to a religious cult. Cheese making was a common dairy practice. Depictions of Minoan ships had been

LADIES IN BLUE *wear elaborate hairstyles and breast-baring bodices fashionable among high-born Minoan women (above) in a wall painting at Knossos. Death-defying athletes somersault over a charging bull in another fresco (above left); bull jumping was a favorite spectacle of the Minoans. The sport may have inspired the Cretan myth of the Minotaur.*

unearthed since Evans began work on Crete, but their quality and detail were poor. A miniature frieze immortalized a waterborne race or procession. This naval scene of a flotilla of ships offered the first glimpse of what Minoan vessels looked like. Their bows were embellished with birds, flowers, and butterflies, and tapered to needle-shaped points; leaping felines adorned their sterns. One scene portrays ships with bodies floating in the water and a line of armed men moving inland.

Marinatos was killed in a fall in 1974 while at his excavations in Akrotiri. Other archaeologists have continued adding to our knowledge of the Minoans, and still more continue to explore the mysterious downfall of this Bronze Age civilization.

In 1962 Nicolas Platon discovered a fifth and unplundered Minoan palace at Kato Zakro, Crete. John and Effie Sakellarakis unearthed two intact burials at nearby Arkhanes in 1975. The first earthquake victims were excavated a few miles away, at Anemospilia, in 1979. About ten years ago, on the east coast of Crete at Palaikastro, archaeologists found a finely sculpted, 20-inch-high statue of a young runner. This piece suggests that Minoan artists were ahead of the realistic art styles of the Greeks by some thousand years.

Archaeologists discovered the palace-center of Mallia, on the seacoast east of Knossos, in 1915. A thriving seaport, Mallia could have been swept away by great tidal waves set in motion by the eruption. However, villages of the rich scattered across the island, many far beyond reach of even the most powerful of tsunamis, were destroyed as well. Minoan palaces did not fall at the same time, which implies that earthquakes were probably not the cause of their demise; the complexes seem to have fallen at various times, most around 1450 B.C. Knossos, the largest and most powerful palace-center, survived somewhat longer. Despite all the calamities that befell Crete after the volcanic eruptions, its population was not wiped out. Minoans continued to occupy farms and villages throughout Crete, though probably in greatly reduced numbers.

Although the eruption of Thera and the destruction of Minoan Crete occurred so close in time and space that historians for years assigned a cause-and-effect relationship to the two events, there are equally convincing arguments that Minoan civilization survived the natural disaster. Little ash has been found in archaeological digs on Crete, and there is no evidence that tsunamis swept inland across the island. Experts now conjecture that no one volcanic cause—eruption, tsunami, earthquake, or ash fall, or even a combination of them—can account for the ruin of the Minoan palaces. Archaeologists are beginning to consider that the actual destruction of the Minoan palaces and villas was the work of humans. The social structure of the Minoans could have come undone after the volcanic eruptions as an indirect result of those disasters. Crete may have begun to suffer symptoms of social strain that led to internal violence. There is evidence of overcrowding. Perhaps increasing wealth and power in the hands of fewer and fewer people led to resentments and rebellion. Although traditional views of the Minoans emphasized their peacefulness, researchers recently have discovered the remains of large Minoan fortifications on Crete that may encourage reinterpretations.

Evidence increasingly points to conquest from overseas as the ultimate cause of the Minoan downfall. Mycenaean tribesmen on the Greek peninsula became more and more aggressive as they grew in numbers and power. For some 200 to 300 years before the collapse of the Minoan palace system, the Minoan and Mycenaean civilizations had growing commercial and cultural contact with one another. Unlike the Minoans, the Mycenaeans were conspicuously warlike. Archaeological finds at Knossos—weapons, pottery, and clay tablets, written in an early Greek script known as Linear B—prove a Mycenaean presence there after 1450 B.C. It appears that Knossos survived because the Mycenaeans wanted to use the palace themselves. About 1380 B.C. Knossos, too, was destroyed, presenting yet another mystery. With the fall of Knossos the long history of the Bronze Age palatial society on Crete ends. Who brought the final destruction to Knossos? Why? Did Cretans rebel against their Mycenaean overlords? Did the Mycenaean invaders battle one another? Historians do not know and may never find out.

They do know that many Minoans survived the collapse of their civilization. The Bible mentions that the Philistines came from Crete. Archaeological evidence suggests that to a large degree the Philistines were the descendants of migrating Mycenaeans. Minoan religion helped shape the gods of the Greeks, and Minoan pottery and jewelry have been unearthed in Mycenaean tombs. Minoan civilization shines brightly in history, comparable in its enduring achievements to that of the ancient Egyptians. Yet the Minoan is still one of the least known of all cultures. What finally ended it? Without the eruption of Thera, would the civilization of the Minoans have survived to outshine that of the Greeks who followed them? ■

SERPENTS WRITHE *in the grasp of a snake goddess unearthed at Knossos (left). The Minoans left a rich legacy of sacred art, but their beliefs—and the symbolism of the 11-inch faience figurine—remain obscure. Gold seal rings bore images of boats (above) as well as goddesses, demons, and women.*

LOST—AND FOUND ATLANTIS AND THERA

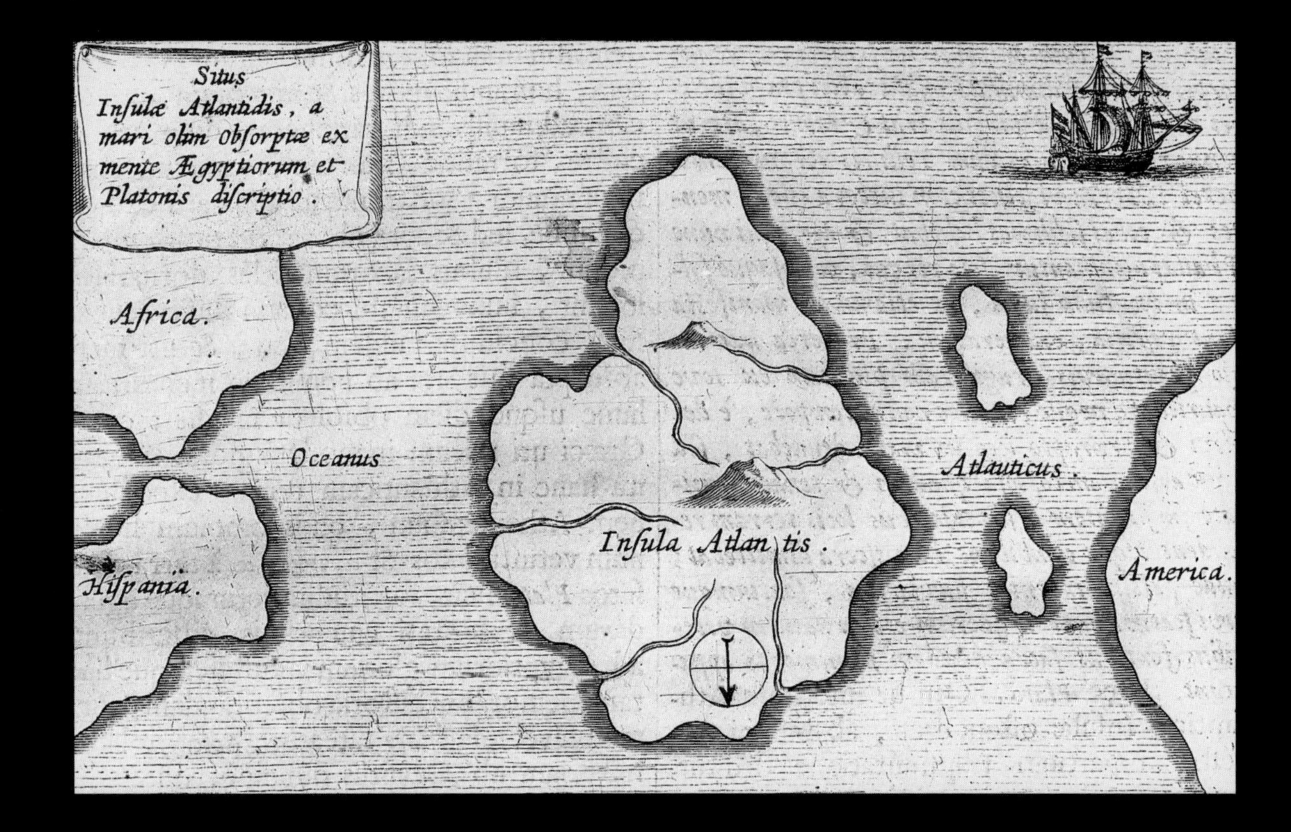

The legend of Atlantis has captivated readers since the time of Plato, yet two of his dialogues remain the only known written records that refer to that lost civilization.

In the 17th century German Jesuit writer Athanasius Kircher published a book entitled *Mundus Subterraneus (Underwater World)* in which he postulated that the continent of Atlantis lay in the Atlantic Ocean. He even included a map of the continent in his book (above).

French author Jules Verne included a description of Atlantis in his 1870 classic science fiction novel *20,000 Leagues Under the Sea,* in which Captain Nemo travels the sea in his fish-shaped submarine named the *Nautilus.*

An engraving by the illustrator of Verne's book, Alphonse Marie de Neuville, reveals the classic columns and broken pillars of a sunken settlement as the explorer notes: "There, before my eyes appeared a ruined city" (opposite).

Since Sir Arthur Evans discovered the Minoan civilization in 1900, scientists have looked for a cause for the demise of Crete's great culture. Greek archaeologist Spyridon Marinatos was convinced that the eruption of the volcano on Thera, north of Crete, about 1500 B.C., was responsible for the fall of that civilization at its peak.

Prior to the eruption, Thera was a round island with the volcanic cone at its

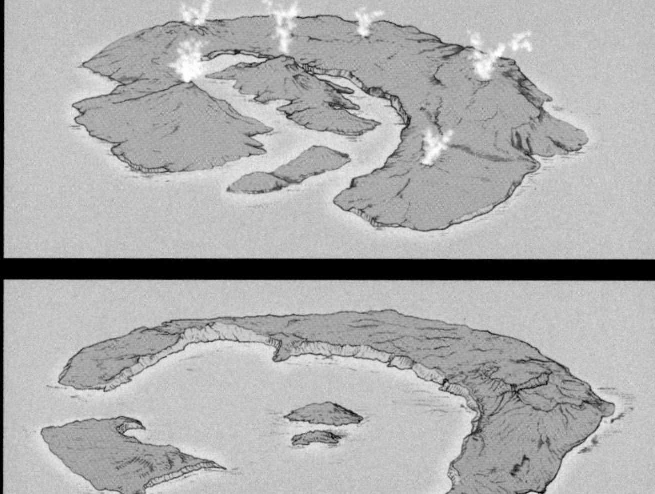

center (top). When the 3,000 foot mountain erupted, the island fell into the sea, leaving five fragments (above).

The sunken caldera of the volcano plunges from 200 feet near shore to more than a thousand feet. The largest of the remnants, Santorini, curves around the colossal subcrater that remains of the original volcano, embracing a seven-mile-wide lagoon. The second largest island Thirasía lies beyond a channel about a mile wide. Two islands in the center of the lagoon—Nea Kameni and Palara Kameni—are the tops of still active volcanoes; the tiny islet of Aspneniu rests below them.

The mystery remains. No one really knows—and we may never be sure—what destroyed the Minoans. Like the legend of Atlantis, however, their legacy survives.

"In the midst of the wine-dark sea," extolled in Homer's Odyssey, a modern vessel (opposite) plies the depths of Santorini's lagoon at sunset.

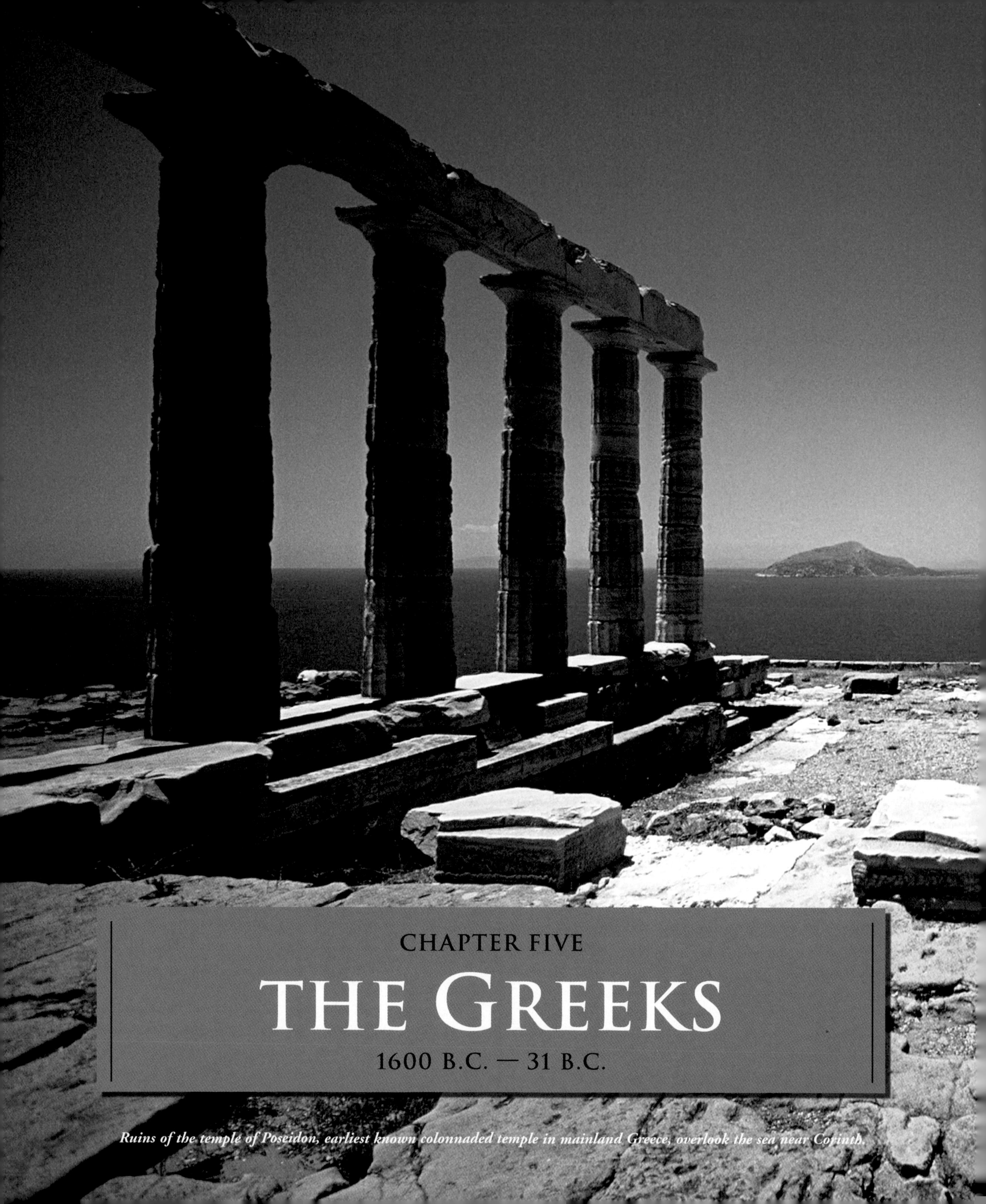

CHAPTER FIVE

THE GREEKS

1600 B.C. — 31 B.C.

Ruins of the temple of Poseidon, earliest known colonnaded temple in mainland Greece, overlook the sea near Corinth.

THE GREEKS OF HOMER

BY ROBERT D. BALLARD

GREEK CIVILIZATION WAS WELL ESTABLISHED AND ITS traditions respected long before Homer immortalized the heroes of the Trojan War in his poems, the *Iliad* and the *Odyssey*. The story of Odysseus's fabled journey—the first great epic of the sea—took place about 1200 B.C. By then the Mycenaeans of the Greek mainland were feared and respected mariners.

The works of Homer united the Greek people and became an integral part of the culture, spread by the Greeks throughout the inhabited world. Epic poetry is the source of much that we know of the Greeks, and excavations on land have contributed a great deal more knowledge. Archaeologists have unearthed pottery manufactured in Greece more than 3,500 years ago from sites in the Near East and Egypt. Painted on vases and incised in stone, many images of the typical Mycenaean single-masted galley survive.

During the eighth century B.C. Greek seafarers traded with Phoenicians in the cities of the Syro-Canaanite coast and picked up knowledge from Semitic shipwrights; products from the Aegean and the Near East have been excavated in Sicily and other western Mediterranean sites. Mass-produced Athenian pottery from the classical period has been found throughout the

Mediterranean Sea. In 1836 workmen in the ancient Athenian port of Piraeus discovered inventory records dating to the fourth century B.C. that hint at the huge size of the city's trading fleet.

Little direct information exists about the maritime merchants who made the rise of Greek civilization possible. Nor is there any information in ancient literature that describes how many ships were lost at sea, an estimate of ten percent is probably low. Considering the tens of thousands of ships that sailed to and from Greece and the Greek colonies, we have found very few. One of the first wrecks that attested to the far-reaching nature of Hellenic trade was found off Greece in 1900. The *Antikythera* proved to be a Roman galley from the first century B.C. loaded with Greek sculptures of bronze and marble. In 1968 a team under the direction of archaeologist Michael Katzev conducted an extensive excavation of a wreck from the fourth-century B.C. found less than a mile from shore near Kyrenia, Cyprus.

The most exciting Greek shipwreck found thus far is also the most recent. In the summer of 1999 George F. Bass and his team from the Institute of Nautical Archaeology at Texas A&M University led an expedition to study a ship that dates from the time of Pericles and Sophocles. It is the first wreck from Greece's golden age—the fifth century B.C. The ship lay 140 feet deep off the Aegean coast of Turkey, between the islands of

Samos on the south and Chios on the north, both part of the Athenian empire. The vessel was modest in size, indicating that it probably performed day trips, and the cargo was not a rich one, implying that the owner may have borrowed money with interest to purchase it. Many amphorae contained organic remains such as pine tar for ship caulking and cattle bones that were originally the "salt and ribs of beef" cited in ancient ship manifests. In addition to amphorae, the team recovered round water jars called *hydriai,* large bowls, and pitchers, all items used to mix wine with water, decant, then serve it. The great number of pottery items found indicates that they were part of the cargo, not for shipboard use. Although wood-borers had devoured much of the ship, the size and shape of a copper nail recovered from the site attest that the boat's builders had nailed an internal frame to a wooden hull held together with mortise-and-tenon joints—a design that was standard in later Greek and Roman ship construction.

With each wreck marine archaeologists find, they broaden our knowledge of ancient maritime civilizations. There are plenty of them out there, and with more sensitive radar, more sophisticated submersibles, and more delicate machinery, we will find them. ■

"THE TERRITORY SHOULD BE NEAR A PORT SO AS TO HAVE ACCESS TO TRADE, BUT NOT RIGHT ON THE SEA SO AS TO AVOID INFLUX OF FOREIGNERS."

PLATO, "CRITIAS"

CAREFUL KEEPING

Artifacts recovered from the deep offer vital information about the civilizations that created them. Conservators treat each object with scrupulous attention and use methods old and new to preserve it and the knowledge it offers. On board the support ship Knorr, *archaeologist Cheryl Ward, giddy with the prospect of researching a new find, grips a 1,500-year-old amphora recovered from a wreck during Robert Ballard's 2003 expedition (opposite); unsealed, the jar contains a residue of its original cargo. Technology has changed record-keeping as well as exploration. Chief conservator of the 2003 expedition, Dennis Piechota, enters the day's data in his laptop (right top), as he does each evening aboard ship during an expedition. Conservators still use tried-and-true techniques to save each treasure recovered. First, an amphora is carefully rinsed and dusted with a brush to remove sand and saltwater (right). Next, the artifact is placed in a pool of half saltwater and half freshwater to desalinate it, then it's dried slowly.*

ATHLETES IN ART

Sports and physical fitness ranked high in Greek culture. A discus thrower decorates a red-figured cup that dates to the sixth century B.C.

GREEK CULTURE

MORE THAN 9,000 YEARS AGO, WHILE HUNTERS and gatherers roamed Greece and long before farmers and shepherds worked its hillsides, its sailors were already exploring the Aegean Sea. In Franchthi Cave on the Peloponnesus, the peninsula that forms the southern half of the Greek mainland, archaeologists have excavated obsidian blades imported from the island of Melos, in the Cyclades. By the beginning of the Bronze Age, about 3000 B.C., Greek farmers were cultivating olives and grapes. Yet the height of the Greek classical period—the culture that bequeathed to the Western world the law codes, government politics, styles of architecture, fields of study, literary works, and vocabulary that are integral to our daily lives—lasted only from the fifth to the second century B.C. Even our alphabet is a legacy from the Greeks; they added five Semitic consonants to be used as vowels in the system of characters inherited from the Phoenicians.

From the very beginning of their history, the Greeks looked to the sea. Much of their homeland consisted of poor soil and barren, forbidding mountains. Most people lived near the coasts, and they naturally gravitated to the waters that surrounded the Greek mainland—the Ionian and the Aegean Seas. By 200 B.C. so many hundreds of their city-states hugged the Mediterranean that Athenian philosopher Socrates had likened them to "frogs around a frog pond." The sea served not only as a highway for the exchange of goods and supplies but also as an avenue for the spread of Hellenistic culture.

By 1600 B.C. the Mycenaeans, or Achaeans—aggressive tribesman who moved south from Macedonia—had settled in mainland Greece. These Bronze-Age Greeks traded with the Minoans, imitated their culture, and finally replaced them. Archaeological evidence suggests that some 150 years after the volcano erupted on Thera and devasted Crete and its Minoan culture, the Mycenaeans overran the island about 1450 B.C. After that the Mycenaean became the predominant Aegean culture.

Elite soldiers who fought in two-man, horse-drawn chariots—fierce Mycenaean warlords bearing armor and weapons of bronze—created small chiefdoms on the Greek mainland and soon asserted their power beyond it, establishing colonies in Italy, the islands of the Aegean, and along the west coast of modern Turkey. Wealth from trade created class distinctions and the need for defense. The warrior aristocracy that later inspired the epics of Greek poet Homer headed a complex social, economic, and political structure. The Mycenaeans amassed vast wealth, and they did not hesitate to stoop to piracy to further enrich their coffers. Their shaft-shaped tombs contained treasures of gold funerary masks, swords, and jewels.

The Mycenaeans built heavily fortified, palace-oriented cities with a commanding view of the surrounding farmlands. Many of the palaces dotted the mainland with names such as Pylos, Tyrins, and Gla. Mycenae, the chief stronghold and the one from which the civilization derived its name, overlooked a key trade route in the eastern Peloponnesus. A single gate permitted entry through its 26-foot-high walls. According to the *Iliad,* in about 1200 B.C. warlords led by Agamemnon, the king of Mycenae, united against the city of Troy, on the Hellespont in Asia

ROCKY CITADEL

of Mycenae, main stronghold of the ancient Greeks, topped a hillside commanding the Gulf of Argos. The stone-walled cemetery at center—Grave Circle A—held a cluster of six shaft graves.

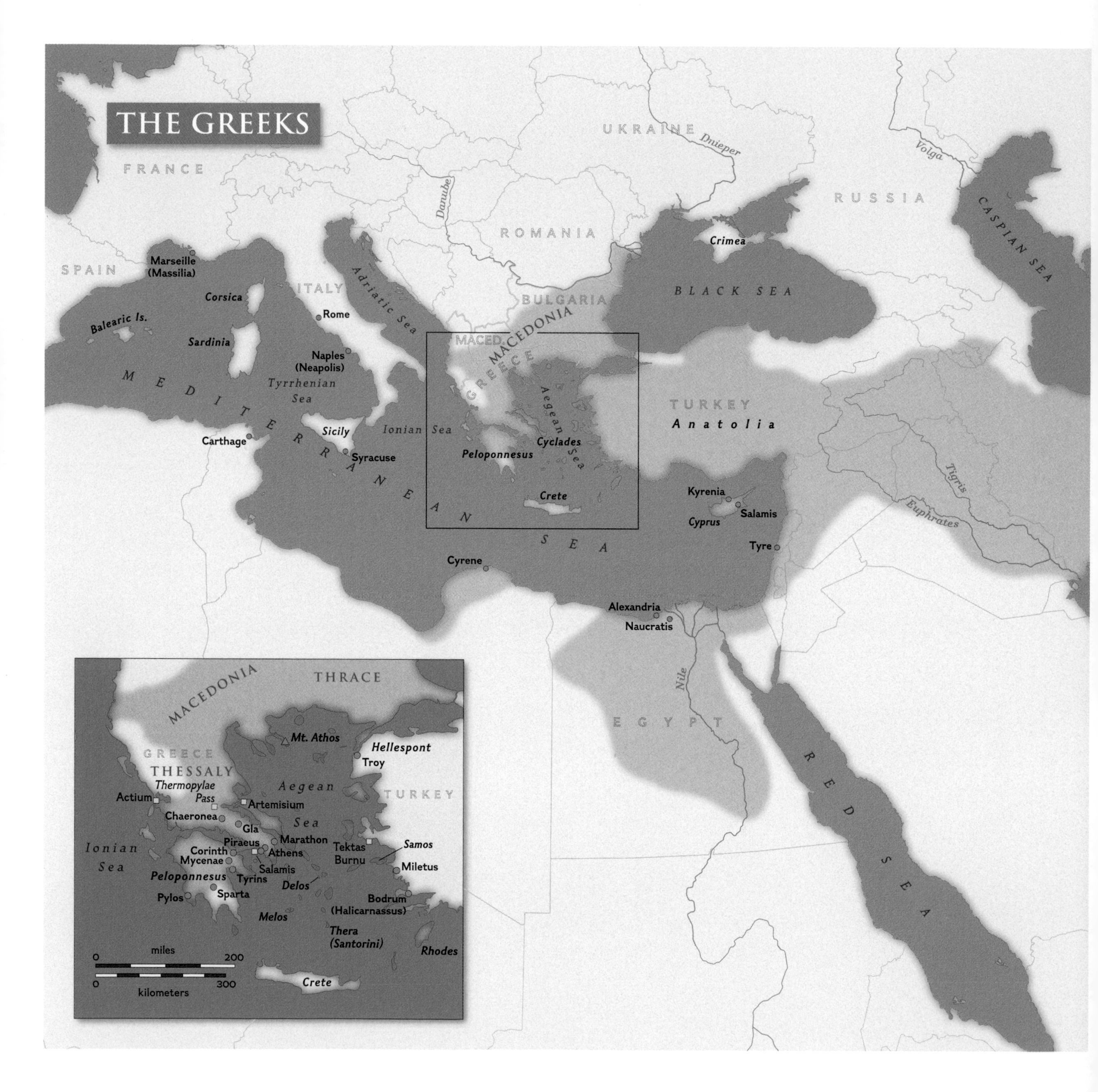
THE GREEKS
FRANCE
SPAIN
ITALY
UKRAINE
RUSSIA
ROMANIA
BULGARIA
MACEDONIA
MACED.
GREECE
TURKEY
Anatolia
EGYPT
Dnieper
Volga
Danube
Tigris
Euphrates
Nile
CASPIAN SEA
BLACK SEA
Crimea
Marseille (Massilia)
Corsica
Balearic Is.
Sardinia
Rome
Adriatic Sea
Naples (Neapolis)
Tyrrhenian Sea
MEDITERRANEAN SEA
Sicily
Syracuse
Carthage
Ionian Sea
Aegean Sea
Cyclades
Peloponnesus
Crete
Kyrenia
Salamis
Cyprus
Tyre
Cyrene
Alexandria
Naucratis
RED SEA
THRACE
Mt. Athos
Hellespont
Troy
THESSALY
Thermopylae Pass
Actium
Artemisium
Chaeronea
Gla
Marathon
Piraeus
Corinth
Athens
Mycenae
Salamis
Tyrins
Delos
Tektas Burnu
Samos
Miletus
Peloponnesus
Pylos
Sparta
Melos
Bodrum (Halicarnassus)
Thera (Santorini)
Rhodes
Crete
0 miles 200
0 kilometers 300

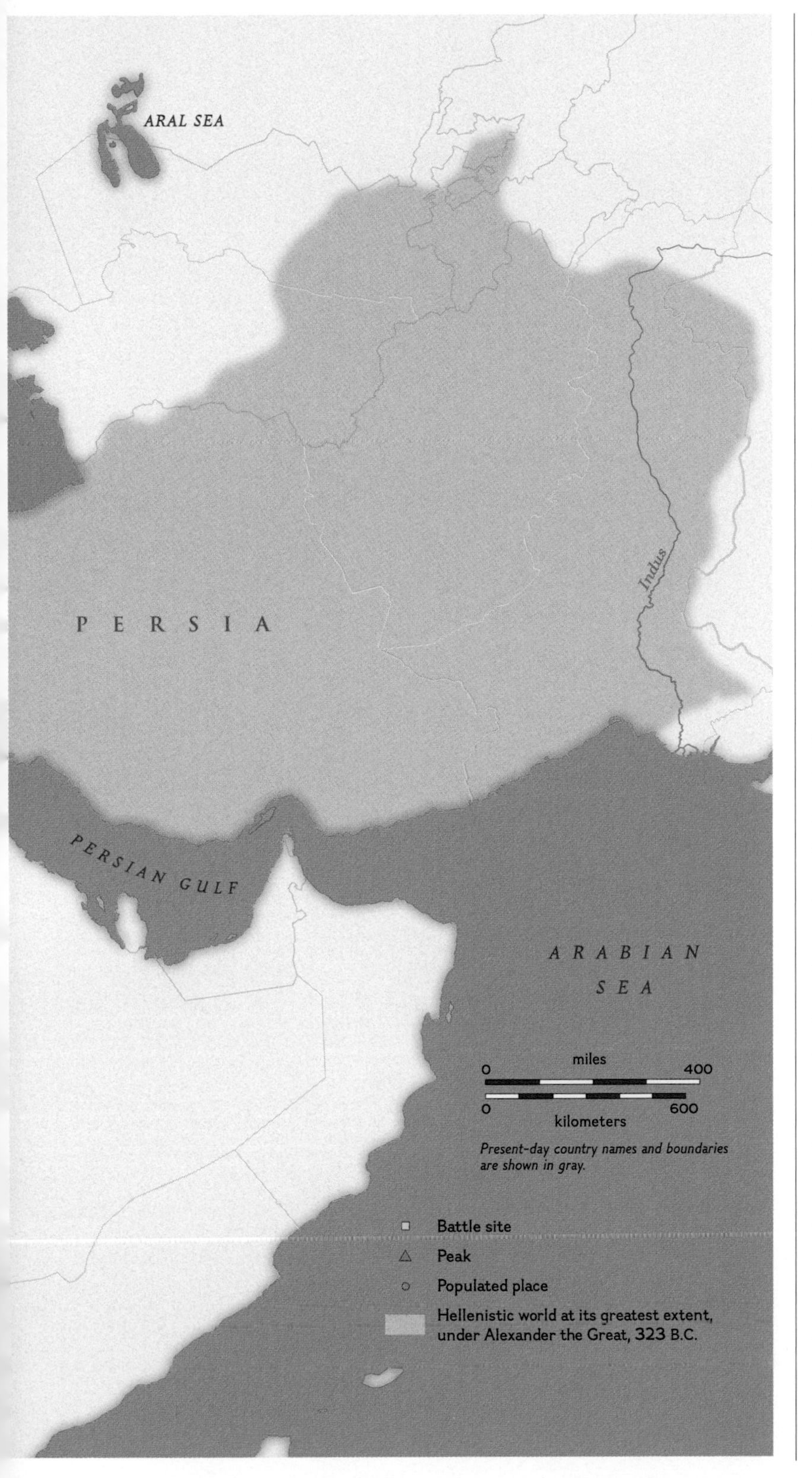

Minor. Myth and fact seem to meld in Homer's epics, which are the result of a lengthy period of oral transfer, during which these stories were relayed and embellished by traveling bards. In 1876 German business magnate and amateur archaeologist Heinrich Schliemann, whose earlier discovery of the Anatolian ruins heralded as Troy had made him famous, unearthed rich burial gifts at Mycenae that attested to the wealth of the warrior elite as well as the skills of Greek craftsmen.

At the height of their prosperity, Mycenaean kingdoms traded across the Mediterranean to Sardinia, southern Italy, and Sicily, east to Troy in present-day Turkey, and north into Macedonia. The economy grew to depend on this far-flung trading network and on concentrated grain cultivation. Mycenaean palaces imported copper and tin, which they needed to craft weapons, armor, and tools of bronze as well as jewelry and personal items. Organizing commerce, maintaining an adequate food supply, and assigning teams of men to work and train for war necessitated that the Mycenaean bureaucracy be efficient. However, it was also vulnerable and sensitive to breakdowns in communications or poor harvests or a commercial failure.

Historians are not sure what may have been the cause, but toward the end of the 13th century B.C., disasters caused the collapse of the Mycenaean palace-based society. Perhaps the climate changed, bringing famine; maybe invaders from the north reduced the hilltop palaces to rubble and ashes. Major trade ceased; fortified Mycenaean cities disappeared, and Greece was depopulated. The former palatial centers were now replaced by isolated farming communities.

Five centuries of economic depression, isolation, and social unrest—a dark age—followed the collapse of the Mycenaean empire, but this period held the seeds of the so-called classical culture that would one day spread across Europe and the world.

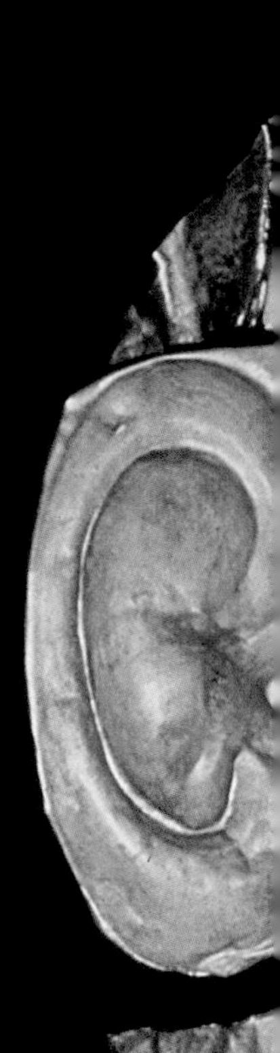

GRAVEYARD GOLD: *Five of the royals buried in the shaft graves of Mycenae's Circle A wore funeral masks. In 1876, when Heinrich Schliemann discovered it, he thought the 3,500-year-old mask hammered from a sheet of gold (right) was that of Mycenaean King Agamemnon; in fact, it predates him. A two-handled chalice (above) was named after Nestor, wise counselor in Homer's* Iliad. *Too large for the living, a golden diadem (below) measured more than 25 inches in width.*

Homer

Socrates

Aristotle

Greek culture experienced a rebirth beginning in the eighth century B.C. as towns and farming communities gave rise to city-states on the mainland, and settlements along the coasts of the eastern Mediterranean, and in Sicily and southern Italy. Trade and sea travel revived. During this time, known as the Archaic Period, each city-state, or *polis,* consisted of an urban center with rural territory around it. Many city-states built a fortified *acropolis,* or "summit city," on a height above the town. Their love of freedom and independence kept the Greeks fragmented; each city-state existed independently of the others. Kings ruled at first, then self-appointed tyrants held sway in many city-states. Later, an aristocratic ruling class took control, and the citizens of many city-states gradually acquired a voice in their own government—the basis of the democratic government we admire today. Most city-states numbered fewer than 20,000 enfranchised citizens, but many were so large for the times that they could not support their blossoming populations.

To control their own numbers, during the two centuries beginning about 750 B.C. the Greeks sent excess population to colonize the coasts of the Mediterranean and the Black Seas. Usually fewer than 200 emigrants, backed by sponsors in their home city-state and blessed by the oracle at Delphi, set off to launch a new settlement. Three branches of peoples made up the Greek world—Aeolians in northwestern and central Greece, Ionians in Attica and western Asia Minor, and Dorians of the Peloponnesus, Crete, Thera, and southern Asia Minor, including the island of Rhodes. Each city or region developed its own character: Thessaly tended toward rural conservatives; the people of Thebes were adept in business; Spartans were soldiers; and the inhabitants of Corinth excelled at commerce.

Their political systems also came to differentiate the city-states. Corinth was an oligarchy ruled by the wealthy; Sparta was controlled by pairs of kings; and Athens was managed by popular assembly. No matter their origin or

Plato *Solon* *Pythagoras*

where they settled, however, the Greeks considered themselves one people—a people apart from all others who were foreigners, or "barbarians." Each new colony was linked to its mother city, or *metropolis.* Their language and the literary tradition of Homer also bound the Greeks together wherever they established themselves. They were united, too, by their possession of a common religion that centered around an Olympian family of anthropomorphic gods, headed by Zeus. Their keen awareness of the worth of the individual and their unshakable belief in what each could accomplish set the Greeks apart from any other people of the age.

Although the scope of Greek civilization was vast, many of the most important and powerful city-states lay in the relatively compact area around the Gulf of Corinth. In the eighth century B.C., the trading city of Corinth, which controlled the isthmus between Attica, the barren and mountainous area jutting into the Aegean to the northeast, and the Peloponnesus, rose to prominence. As Greece grew more powerful, its inhabitants built up their navies. Corinthians built the first Greek triremes—swift galleys with three tiers of rowers—and in time maintained four fleets of trading vessels, some dedicated to sailing west, others to voyages eastward.

Increased commerce between city-states, settlements, and barbarians to the east during the seventh century B.C. inspired the introduction of coinage, about 600 B.C. The *hoplitai,* a heavily armed infantry of middle-class soldiers, replaced the aristocratic cavalry of preceding centuries. In addition, a mercantile middle class arose in Greece that blurred the class distinctions between nobles and commoners. This development encouraged the spread of political rights and set the stage for the growth of democracy. For centuries most Aegean cities had been ruled by groups of landed aristocrats. These oligarchs tended to run the governments to their own advantage. They amassed more riches and power as rising numbers of commoners grew discontented. In city after city,

ambitious leaders took advantage of the social unrest and seized power, evicting the oligarchs and installing themselves as tyrants. Although some tyrants were repressive, many sought to rule fairly and imposed much needed political stability.

By the sixth century B.C. the *polis* of Sparta, its traditions of self-sacrifice and military strength integral to its growth, became the leading city on mainland Greece and controlled most of the Peloponnesus. To consolidate its own power, Sparta began to depose rulers of island-states to the east and install governors sympathetic to its goals. Gradually Sparta came to be known as a champion of liberty against tyrannical rule, while still enslaving her local population.

The city-state of Athens suffered the same social ills as its neighboring city-states: a firmly entrenched ruling aristocracy, a growing population of commoners who had no say in government, and rising numbers of inhabitants with no rights who lived in abject poverty. The Athenians, known for their strong-mindedness, energy, and restlessness, turned away from tyranny. In 594 B.C. they named a poet, politician, and general—Solon—as *archon,* the state's highest official. Solon drew up a new constitution, canceled or reduced all debts, and freed all debtor slaves. He offered citizenship to all classes of free men, including foreign artisans, and legally defined the existing class structure of Athens in terms of property ownership and annual income. All male citizens, even those belonging to the poorest, landless class, could sit in the assembly and appeal to the law courts for justice.

The stony hillsides of Attica could not produce enough food to feed the growing population of Athens.

GREEK YOUTH *sprint for the finish in a close-up of an amphora that served as a first prize. The Panathenian Games, precursors of the Olympics, hosted Greece's best athletes for more than a millennium. The winners received amphorae filled with olive oil.*

Solon pushed the city into commerce on a greater scale. He banned the exportation of all agricultural products save olives and oil to pay for the importation of necessary grain from the Nile Valley and the Black Sea. From the Persian Gulf and the Phoenician cities of Tyre and Carthage, Athens bought luxuries such as ivory, frankincense, carpets, and purple dye. The city grew in power and might. Despite Solon's efforts, when he retired from public office, power reverted to the noble families of Athens. Soon quarreling factions brought government to a standstill.

An ambitious nobleman named Peisistratus took advantage of the chaos to seize power by force. For the next 20 years he ruled Athens with energy and skill, transforming it into the leading city of the Greek world. Peisistratus catered to the poor, extending loans to farmers, confirming the rights of artisans, and promoting industry and commerce. Under his direction, Athens became a center of culture and art. New themes in pottery appeared, a result of the influence of trade with the East. Artists increasingly depicted humans and everyday life on vases and began to sculpt monumental figures. Architects designed huge temples in stone, some in the Doric mode with massive fluted columns, later others with the delicate columns on molded bases characteristic of the Ionic style. As the Persian Empire expanded and threatened Greek colonies in the east, refugees from the Ionian coast—among them poets, philosophers, artists, and craftsmen—flocked to Athens.

As Athens bloomed, so did most of the Hellenic world. By 500 B.C. some 250 Greek city-states were scattered from eastern Spain and northern Africa through France and Cyprus, to the present-day countries of Bulgaria, Romania, Ukraine, Russia, and Turkey on the Black Sea. A Greek merchant town in Egypt, Naucratis, on the Canopic branch of the Nile, flourished by exporting linen and papyrus. Garrisons along the approaches to the Hellespont, the ancient Greek name for the Strait of Dardanelles, protected the main route into the Aegean for merchantmen carrying wheat and fish from the Black Sea. Seaports thrived from Corinth, Sparta, and Athens on the mainland to Miletus in Asia Minor, Marseilles in southern France, Syracuse in Sicily, and Neapolis (Naples) in southern Italy. The use of metal coinage supplanted barter, the ancient system of trade between farm and city. Coins of many realms changed hands, from the gold disks of Persia to the silver coins of individual Greek city-states.

The stage was set for the blossoming of one of the most remarkable civilizations the world has known: the classical Greek world that shaped today's Western culture. During the next 200 years, Athenian statesmen transformed the world's political organization; Greek generals effected extraordinary victories; poets, architects, painters, and sculptors revolutionized the arts and literature; sciences and philosophies developed, and Greek thinkers articulated ideas that would change the way human beings view nature and their own existence.

With the expansion of the Persian Empire to the east, trouble began to brew for Greek colonies. The Persian conquest, led by Darius I of the Ionian Greek cities in Asia Minor (in modern Turkey), brought Athens into direct confrontation with the Persians. Athens dispatched 20 ships full of soldiers to help the eastern Greeks rebel against the Persian forces. Darius burned the southernmost city of Miletus with its important harbors to the ground and killed, enslaved, or deported its citizens.

In 492 B.C. a fleet sent by Darius sank during a storm off Mt. Athos, with the loss of nearly 300 ships. By 490 B.C. Athenian forces—blessed with good luck and a wily general named Miltiades—triumphed over the numerically superior Persian forces in the plains at Marathon. The Greeks historically had battled on land and were surprisingly lacking in naval strength. They used ships for commerce and mainly as troop transport. When faced with the Persian threat, however, Athens

HOMER'S GREAT EPIC THE ODYSSEY

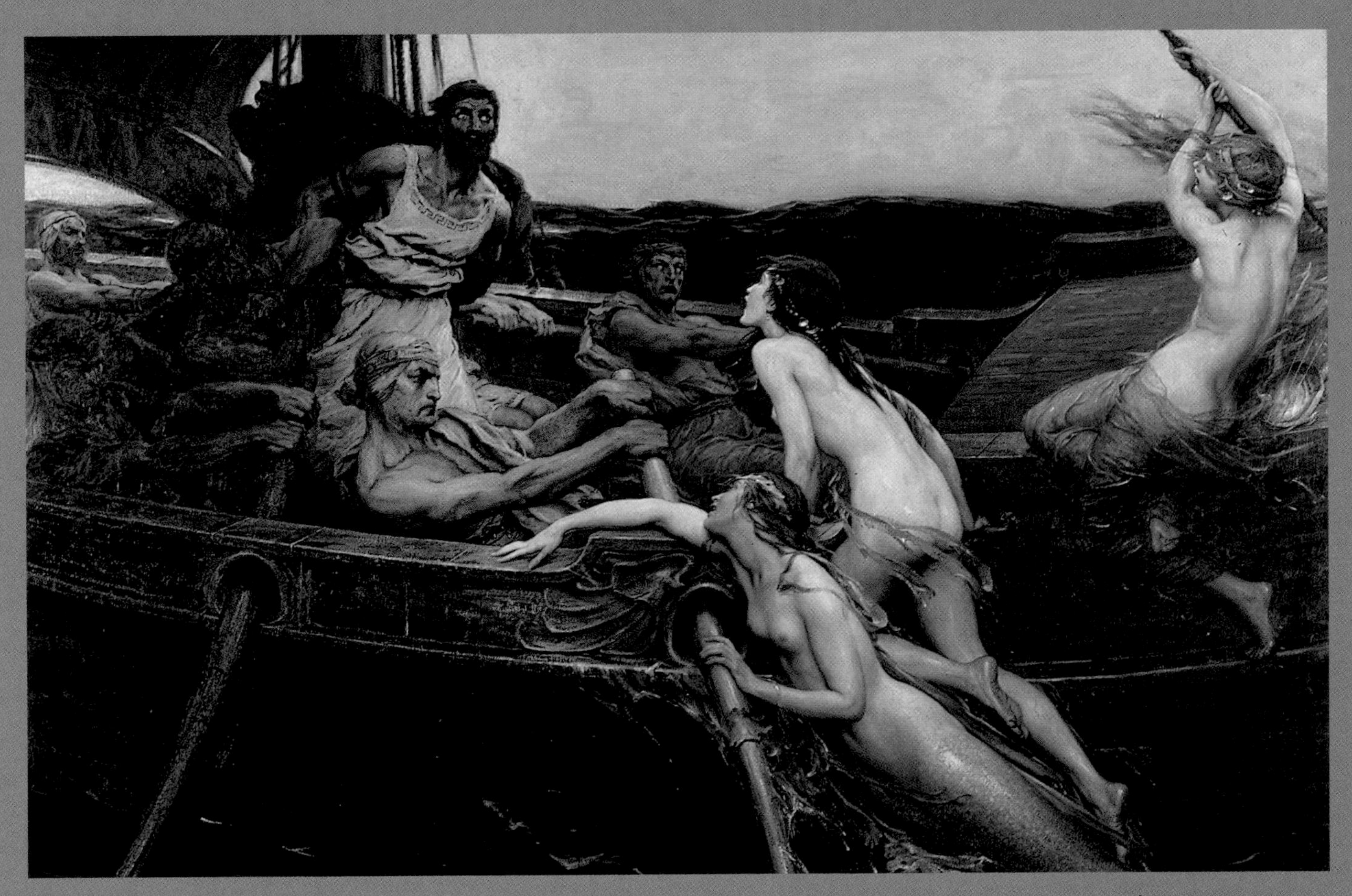

Odysseus struggles against his bonds as his crew, their ears plugged with wax, strain past the Sirens, in an oil painting by H. J. Draper.

Homer's *Odyssey* was written around 800 B.C., but the story takes place some 400 years before, after the ten-year-long battle of Troy, when Greek warlords sailed to Asia Minor to avenge the theft of the beautiful Helen.

After their victory, the most famous of all ancient mariners, Odysseus, along with other Greek chieftains, set sail for home only to find that his voyage was as dangerous and as long as the war itself. Athena and Poseidon, angry at what they considered the thanklessness of the Greek victors, sent a fearful storm that lashed their ships, tearing many of them to bits.

The vessel carrying Odysseus and his crew washed up in the land of the Lotus-Eaters, where he discovered that whoever partook of a strange plant that grew there wanted to stay forever. Odysseus dragged his men on board ship and chained them there while he sailed for safety. At the next stop, the one-eyed goat herder, Cyclops Polyphemus, imprisoned Odysseus and his shipmates in a cave and proceeded to eat them one by one. The giant blocked the cave with a boulder so big only he could move it. To escape his clutches, the Greeks drove a smoldering stake into his eye and rode out of the cave, clinging to the undersides of his goats.

Even one of Odysseus's few happy encounters turned out badly. Aeolus, King of the Winds, gathered all the world's storms into a leather bag and presented it to Odysseus. Mistaking the contents for treasure, his crew opened the sack, and a fierce hurricane blew the ship backward. When the men began to land in the country of the Laestrygons, the giants bashed most of the Greeks' vessels with boulders and ate a few of the sailors before the rest could escape. The survivors found refuge on the island of Aeaca, the realm of the beautiful and dangerous Circe, only to discover that she had transformed them into swine with a flick of her magic wand. Hermes, messenger of the gods, gave Odysseus a divine herb so that he could withstand Circe's magic. She, however, fell in love with the valiant warrior from Ithaca and turned his crew back into men. They tarried a year on her island, and she used her sorcery to foretell their next move. They must, she advised, cross the river Ocean and beach their ship on the shore belonging to the Queen of the Underworld, Persephone. There, at the entrance to Hades, Odysseus should find the ghost of the blind prophet Tiresias, who would tell him how to get home to Ithaca. To attract the shades of hell, Odysseus and his crew filled a pit with lamb's blood and watched a procession of the dead file past to drink. Odysseus saw and spoke to some of the fallen heroes of the Trojan War, among them Achilles and Ajax, before he finally recognized the seer who could help him home.

Warned by Circe, Odysseus plugged his men's ears with wax and had them tie him to the mast to escape being lured ashore by the bewitching songs of the Sirens who would bash their ships on the rocks of their island. Then the Greeks navigated a narrow passage between Charybdis, a foaming whirlpool that swallowed the sea and spewed it forth again, and a gigantic cliff to which was rooted the dog-headed monster, Scylla. Six sailors died as she grabbed them and wolfed them down.

Star of a fifth century B.C. oenochoe, *or wine pitcher, Odysseus, sword in hand, pursues the magical Circe.*

Next, becalmed on the island of the sun, Odysseus's men foolishly sacrificed his cattle. Furious at their temerity, Zeus sent a thunderbolt that smashed their ship to bits. Only Odysseus survived, crashing on the island of the enchantress Calypso. There he settled until the gods intervened, urging him to build a raft and continue his journey. Another storm blew it apart like thistledown, and he finally washed ashore among the Phaeacians, a kind people and splendid sailors. At a banquet in his honor, Odysseus recounted the perils of his long voyage. The king, Alcinous, moved by the tale and impressed by Odysseus's long travails, sent him home to Ithaca.

With Athena's help, Odysseus overpowered the ardent suitors who had been pursuing his wife, Penelope, during his long absence. After a journey so long and perilous that it inspired the word "odyssey," the dauntless Greek mariner finally regained his family and his home.

began to build a fighting navy. The city organized and led a defensive coalition of city-states called the Delian League, centered on the sacred island of Delos—said to be the birthplace of the god Apollo—that shared the expense of maintaining a powerful military fleet. Sparta, through an alliance called the Peloponnesian League, was the strongest land power.

In 480 B.C. the Greek city-states united against the Persian army led by King Xerxes. The Greek army, led by King Leonidas of Sparta, blocked Xerxes's forces at the pass of Thermopylae until a traitor betrayed the Greeks to the Persians. Leonidas, with 300 countrymen, heroically held the pass, reduced to fighting with fists and feet until every man had fallen, so that the rest of the Greek army could safely withdraw and a warning could be sent to the Greek navy. The Greek fleet of some 324 triremes and nine *pentekonters* (50-oared ships)—more than half of them from Athens—fought the Persians to a standoff at Artemisium.

Off the island of Salamis, the Greek navy reassembled. Like the Persian ships, the Greek triremes were galleys that achieved unprecedented speeds of ten knots using three banks of oarsmen. Each was about 115 to 120 feet long, 16 feet wide, and weighed less than 40 tons. Although rigged with two square sails, the trireme relied on oar power during battle and was armed with a bronze-sheathed ram. The Greeks, though outnumbered three to one, enticed the Persian fleet into a narrow channel. As the Persian ships funneled through the opening, Greek triremes rammed them one by one. At Salamis, one of history's great naval battles, the Persians faced the most disastrous defeat in the history of their empire. They withdrew from Athens, never to return; the battle marked the beginning of Athens' golden age.

Three issues dominated the centuries of classical Greek civilization. One was Greek relations with Persia. Another was the intense rivalry between Athens and Sparta, which led to the Peloponnesian Wars. The third was the unprecedented intellectual, political, and artistic achievement of Athens. After the battle of Salamis, Athens's overwhelming naval power enabled it to dominate the Delian League and use it as the foundation of a commercial empire.

In the fourth century B.C. the population of Athens, estimated conservatively at about 150,000 people, survived chiefly on grain imported from the shores of the Black Sea, Egypt, and Sicily, augmented with salt fish from Spain and the Black Sea. Athens was the largest importer of grain at that time. The island of Rhodes, which lay halfway between Egypt and the Crimea, became the cornerstone of the international grain trade. Deep and beamy sailing freighters called *holkades* relied on a broad, low mainsail rather than oars. They sailed mostly between May and October, when the winds and weather were most favorable.

Athens exported olive oil, which served the uses of modern butter, soap, and electric lighting. Pottery played a major economic as well as artistic role. The ceramic wares were mass-produced for export as far as the Black Sea and Africa, and the *Keramikos,* a particular quarter of Athens, was devoted to potters and their workshops. From Macedonia and Thrace, Athens imported timber, and from Egypt, flax. Cyprus contributed copper and Asia Minor, iron. In addition to the critical grain and fish, the Black Sea area supplied hides, cattle, hemp, wax, and slaves, who performed most of the labor in the Greek world. Most Athenians lived simply. They dwelt in modest mud-brick homes and ate meat—usually goat or sheep—only on feast days. The upper class could afford luxuries imported from throughout the world—incense from Arabia, amulets and scarabs from Egypt, Aegean marble, and pigs and cheese from Sicily.

As is true of other civilizations, until fairly recently evidence of early Greek shipping consisted mostly of historical and literary references and ancient depictions of ships.

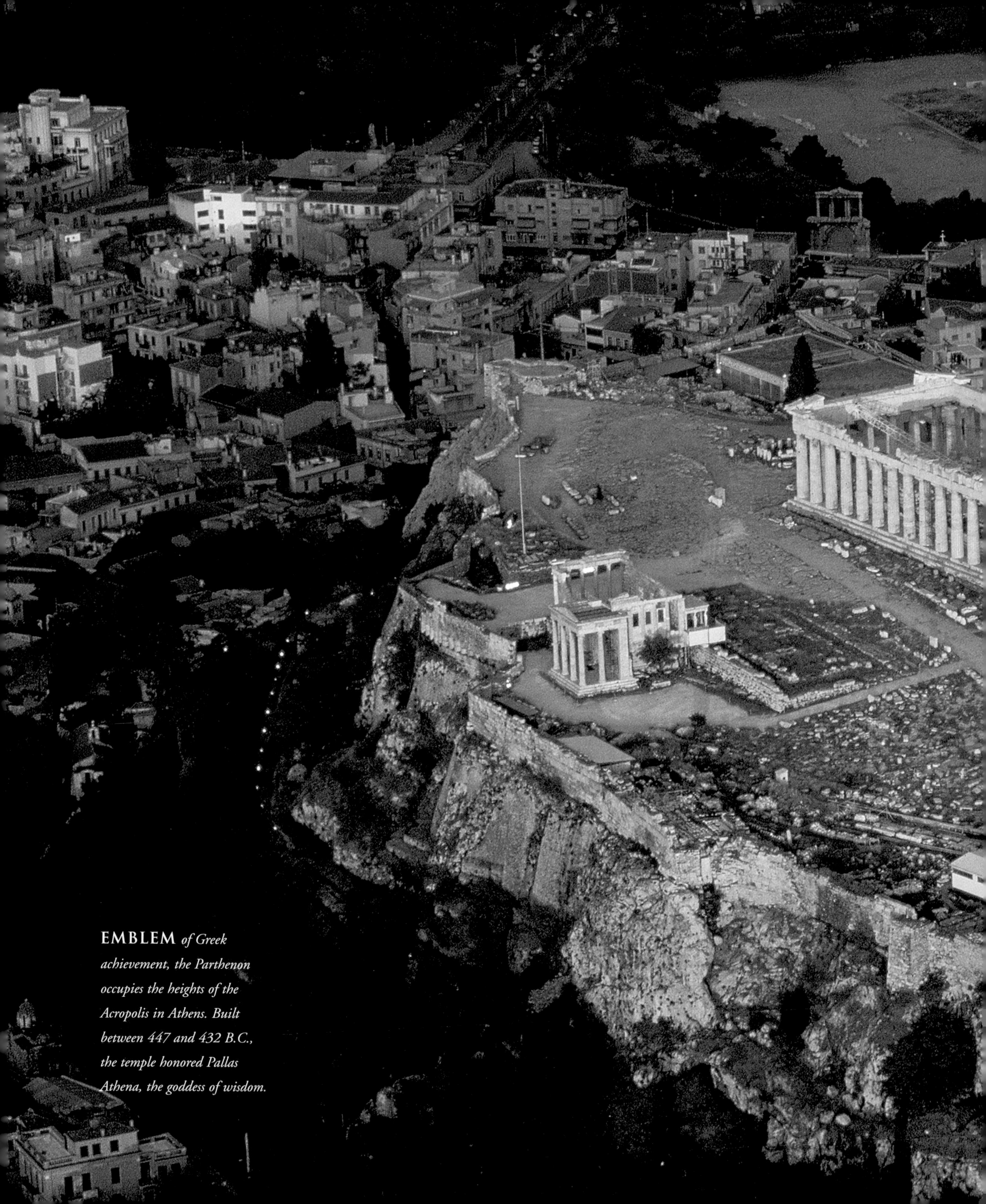

EMBLEM *of Greek achievement, the Parthenon occupies the heights of the Acropolis in Athens. Built between 447 and 432 B.C., the temple honored Pallas Athena, the goddess of wisdom.*

HOOVER

In 1967 a wreck discovered off Kyrenia, Cyprus, suggested that merchant seamen traded a wide variety of items. The expedition, led by Michael Katzev, retrieved ten distinct varieties of amphorae, indicating different ports of call and various commercial commodities, including a great quantity of almonds. Found less than mile from shore, the ship may have been scuttled by pirates near the end of the fourth century B.C.

George F. Bass and his team from the Institute of Nautical Archaeology, based at Texas A&M University, excavated a Greek ship found off the Turkish coast at Tektas Burnu beginning in the summer of 1999. Like the Kyrenia wreck, it sank so close to shore—within a hundred yards—that the crew may have been able to swim to shore. Amphorae and other items recovered indicate that the wreck sank during the time of the Golden Age of Athens, and future study of the relics will add knowledge to the picture of Greek civilization at its cultural zenith.

All Athenian commerce flowed through the triple-harbor complex at Piraeus, which was connected to Athens by a six-mile-long walled passageway. At the port *holkades* juggled for berths along long quays, grain dealers supervised storage in huge warehouses, money changers cried out the latest exchange rates for various currencies, and merchants haggled in many languages in the long colonnade and at the slave market. Maritime trade was risky, and voyages were time-consuming. Most shipments were fairly small, and traders worked primarily from borrowed funds obtained from numerous investors. Foreigners almost monopolized shipping. By law, only Athenians could own real estate—a much safer investment than shipping—so foreign residents in the city were often involved in sea trade. The wharves and docks of Piraeus rang with the voices of owners from Massilia—modern-day Marseilles, the Crimea, and Asia Minor. Pericles, the leader who guided Athens during its apogee, boasted that "the wares of the world find their way to us."

Pericles, an orator and general from an old and noble Athenian family, became the city's leading statesman in about 460 B.C. Intent on giving more power to the lower classes, Pericles removed all economic barriers to public office. Although slaves (about one-quarter of the city's population), women, and foreign residents still had no rights, every Athenian male over the age of 18 could be a member of the people's assembly, the city's most powerful governing body, voting his opinion and proposing measures. The office of *archon,* once available only to members of the upper classes, was opened to a larger portion of the populace. It was the duty of each citizen, Pericles believed, to participate in government affairs, and for the first time citizens were paid to perform civil duties such as serving on juries. Athenians' sense of civic pride and purpose grew as members were selected by lot to a council of 500, the city's steering committee. "We are called a democracy," Pericles explained, "for the administration is in the hands of the many and not the few."

In 449 B.C. Pericles negotiated a peace agreement with the Persians. A 30-year treaty with Sparta followed in 446 B.C. Determined that Athens reflect its standing in the world, Pericles used the time, tributes, and the lavish revenues of the Delian League to rebuild the city destroyed by the Persians before the battle of Salamis. He gathered the finest architects, sculptors, and craftsmen to restore Athens and refurbish the temples and buildings on the Acropolis, the steep hill in the center of the city first settled by Mycenaeans. There Pericles had constructed a massive temple to the goddess Athena, the Parthenon, and leading to it a majestic roadway called the Sacred Way. Below the Acropolis, law courts and public porticoes or porches, called *stoas,* crowded the *agora,* or marketplace.

A new style of art arose during this Golden Age. Artists and sculptors, reveling in the heroes of their Persian victories, began to shape the human body in action,

SOLDIER AND STATESMAN, *Pericles guided Athens into its Golden Age; a bust by Cresilus dates to about 429 B.C. (left). Under reforms Pericles initiated, all Athenian males over the age of 18 voted, casting ballots by scratching their choice on a pottery shard (above).*

relinquishing the stiff forms of former centuries. Artisans flocked to admire the new monuments; with them came philosophers, mathematicians, and writers from throughout the Greek world.

The progressive democratization of Athenian government and society combined with the city's abundance of talent fostered an explosion of artistic and intellectual accomplishments. Throughout the Greek world, people began to question traditional beliefs and practices; artists had more freedom to create. The Greeks loved theater, and each spring three playwrights were invited to compete in a festival to the god Dionysus in Athens. For centuries dramas had consisted of one man interacting with a large chorus. The Greek dramatist Aeschylus, a veteran of the battles of Marathon and Salamis, featured godlike heroes and heroines inspired by the works of Homer. Aeschylus added a second character to his tragedies and reduced the importance of the chorus by stressing dialogue between two protagonists. Sophocles, a politician as well as a playwright, accepted that fate played an important role in existence but maintained that individuals should use their

EVERDAY EVENTS *inspired Greek potters. A tuna merchant discusses a sale on a mixing bowl, or crater, from Sicily (above). In an illustration from a perfume jar (right), a wedding procession approaches the torch-lit house of the groom. Athenian master potter Amasis incised sharp detail before firing the jar in the sixth century B.C.*

free will to govern their lives. His characters, such as Antigone and Oedipus, were ordinary humans determined to stand by their beliefs and fight the whims of fate. Euripides, a contemporary of Sophocles, portrayed starkly realistic characters like Medea and Electra. A fourth luminary of Athenian theater, Aristophanes, found fame writing comedies with political themes such as *Lysistrata,* which promoted his dislike of intellectuals and of politicians who favored war.

Like theater arts, other intellectual pursuits reached new heights during the Golden Age. One of them was the study of history. Herodotus, "the father of history," was born and raised in the Greek colony of Halicarnassus, modern Bodrum, in Turkey. Following civil strife, Herodotus abandoned his home and moved to the island of Samos, but he soon began to travel extensively, interviewing people and gathering notes about their past experiences before settling in Athens. There he continued to chronicle past and present events as objectively and accurately as possible from his vantage point, finally composing a nine-volume history in prose instead of the poetry customary of the time.

The study of medicine advanced through the efforts of Hippocrates. Greeks of the time believed that pain and disease were the work of angry gods or spirits. Hippocrates insisted that health had nothing to do with superstition. The body, he insisted, was a single organism whose parts constantly interacted; a healthy body was one in which the parts functioned well together. Diet and hygiene, he noted, were central to good health.

Philosophy, too, made great strides. Socrates, who taught and questioned in the agora at Athens, urged his students to examine their lives, to know themselves, and to strive for the truth that knowledge discloses. His follower, Plato, would continue the Socratic technique of using questions to pursue knowledge; he founded a school of philosophic and scientific research called the Academy. And Aristotle, one of Plato's pupils, established another school, the Lyceum, where he proposed the use of reason and logic to search for knowledge.

While Athens flourished through its maritime trade and its leadership of the Delian League, neighboring governments became more wary of its aggressive power. The rivalry with Sparta, the city that controlled the Peloponnesus, intensified. Militaristic Sparta, where boys were raised from birth to be foot soldiers in an efficient army, remained supreme on land. By 431 B.C. Sparta and its ally, Corinth, were at war with Athens; it was a struggle

that would last nearly 30 years and would involve virtually the entire Greek world.

Athens entered the Peloponnesian War with the finest navy afloat, but its strength did not assure victory. A plague swept the city, wiping out one-third of its population, including the great statesman Pericles. In 413 B.C., when the Athenian fleet attempted to add Syracuse, in Sicily, to its empire, Syracuse resisted and destroyed nearly two-thirds of Athens's ships and routed the Greek army. Sparta, meanwhile, built its own navy to challenge Athens at sea. When the Athenian fleet, greatly reduced in strength, sailed to the Hellespont to protect the grain shipments that had to pass through en route from the Black Sea, the Spartan ships stealthily followed. As the Athenian ships lay beached and undefended, the Spartans rushed in and destroyed them. Then they laid siege to Athens, destroying the walls that protected the city and led to the port of Piraeus. The Spartans so revered the learning and majesty of Athens, however, that they allowed the city to stand rather than destroy it.

Even this defeat, the greatest Athens ever suffered, did not extinguish its glory. Democracy was reinstated in the city; merchants at Piraeus continued to profit from their far-flung trading network, and scholars such as Plato broadened the Greek's intellectual horizons. Yet power had shifted. The Greek world had changed, setting the stage for even greater changes.

Sparta could not maintain its hold on Greece, and its might faded. In 338 B.C. Philip II of Macedon took control with a victory at Chaeronea, consolidating the Greek world for the first time in history. During the period that followed, which lasted until the Roman conquest in 31 B.C., Philip and his son, Alexander the Great, achieved a unity that the fragmented Greek city-states had never attained. Although the *polis* had offered Greeks the freedom and autonomy that inspired democracy, it had denied them unity and the possibility of nationhood. Now, during the time known as the Hellenistic age, the concept of provincial city-states developed into one of cosmopolitan world citizenship. Alexander, who had been tutored by Aristotle, consolidated his Greek empire and extended Hellenic culture over the inhabited world. Throughout his life, Alexander carried with him a gift from Aristotle—an annotated copy of Homer's *Iliad.* He shipped specimens from exotic lands back to his mentor for study.

Cultures of different races, religions, and traditions lived side by side in the Hellenistic world, but they were unified by the Greek tongue, which became the language of communication and trade. Alexandria, in Egypt, eclipsed Athens as the international center of culture, but learning continued to be cherished and cultivated. Euclid, who established a school of mathematics in Alexandria, wrote a textbook on the principles of geometry, which shaped that science for two centuries. Archimedes, a native of Syracuse, established the physical law of buoyancy and mastered the principle of the fulcrum and lever. Eratosthenes, a native of the Greek city of Cyrene, in northern Africa, who came to head the great library in Alexandria, was the world's first systematic geographer. He calculated the circumference of the Earth within 200 miles of modern measurements. Epicurus, who established a school of philosophy that advised gaining the most possible pleasure while avoiding all possible pain, has followers today. Zeno, who believed the only real good is virtue based on knowledge, advised a life of strict moral discipline. His philosophy,

GREEK TRIREMES *challenge Persian galleys in a 19th-century re-creation of the pivotal battle of Salamis in 480 B.C. Outnumbered three to one, the Greeks maneuvered their enemy into a narrow channel and fought them one by one. The Persian defeat, greatest in its history, established Greek supremacy in the Mediterranean.*

CONQUERING KING: *Philip II of Macedon, sculpted in ivory in a statue from his tomb (opposite), consolidated the Greek world beginning in 338 B.C. His reign initiated the Hellenistic age, which lasted until the Roman conquest in A.D. 31. Cobbled floors and standing columns remain of Pella (above), ancient capital of Macedonia and birthplace of Philip's son, Alexander the Great.*

Stoicism, received its name from the Greek word for porch—*stoa*—where he taught. A new realism in art produced lifelike statues of famous people and masterpieces such as the Winged Victory of Samothrace and the Aphrodite of Melos, or Venus de Milo.

Artists and thinkers of the Hellenistic age preserved the Greek traditions that would form the foundations of Western civilization, but after the death of Alexander in 332 B.C., the struggle for succession caused unrest throughout his empire. And the power of Rome was on the rise. Through the years the Hellenistic states looked more and more to Rome for leadership, as conflict between them escalated. In 148 B.C. Macedonia became a Roman province, and the battle of Actium, in 31 B.C., marked the end of the Hellenistic world. Still, even its conquerors helped preserve its culture: Rome, which admired Greek achievement, imitated its culture and style, art and philosophy, thus helping safeguard the Hellenistic legacy. ■

HISTORY *survives as Alexander charges Persian forces at the Battle of Issos. Little Hellenistic painting survives; a Roman copy of a fourth-century Greek work, this detail comes from a Pompeiian floor mosaic.*

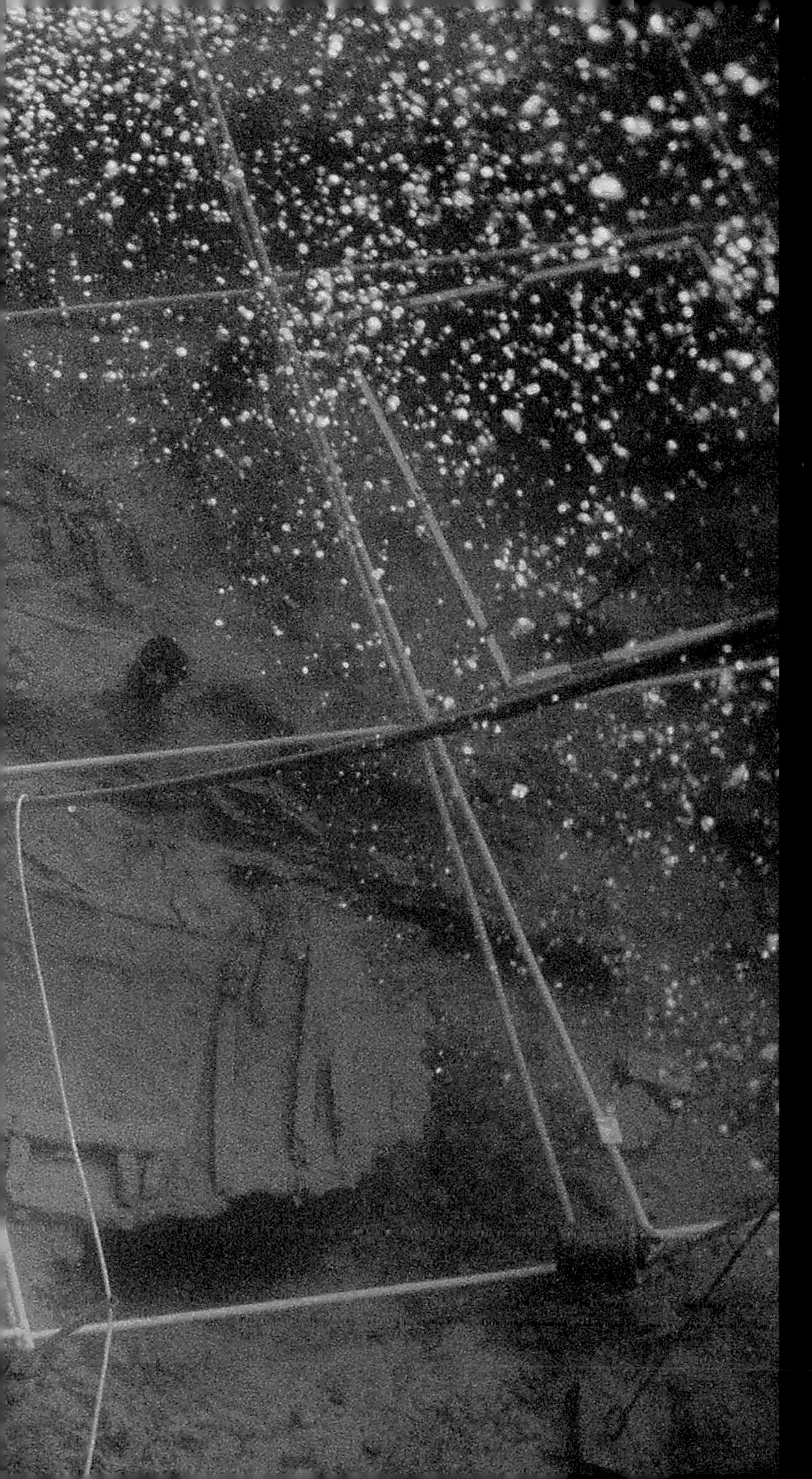

UNDERWATER
ARCHAEOLOGY

A Turkish sponge diver intent on preserving his town's history directed University of Pennsylvania graduate student Michael L. Katzev and his wife, Susan, to a mound of amphorae 90 feet below the sea less than a mile off the port of Kyrenia, Cyprus.

In 1968 they began two seasons of excavation of the oldest known Greek ship yet discovered—a merchantman from the fourth century B.C. that probably foundered in a storm. A scaffolding of plastic pipe divided the site into numbered squares so that each item could be recorded where found (left). Black air hoses fore and aft vacuumed silt as the divers worked. The port side of the vessel, foreground, settled into mud, which helped preserve it from the decay that riddled the more exposed starboard side.

Seaweeds encrusted the mound that hid the Kyrenia ship. David I. Owen designed a pipe with sharp perforations at its tip, which forced compressed air into the sea bottom and loosened the masses of roots in the underwater sod (above).

Beneath the sand a 60-by-30-foot pile of amphorae

each piece of wood—more than a thousand—before it left the seafloor. The markers would simplify reconstructing the ship on land.

Hull planks were loaded onto huge trays tethered to a balloon lift. A diver accompanied each ascent, carefully regulating the air in the balloon to assure that the tray rose slowly. Over a period of two months, the team raised 29 tray loads, more than five tons of waterlogged wood. On shore, workers hosed down the soft and fragile timbers to protect them against warping and shrinking. Then they soaked the timbers with polyethylene glycol, a waxlike preservative that replaced the riddled wood and stabilized it. Carbon-14 dating indicated that the hull planks were cut about 389 B.C. ; the Kyrenia ship had been sailing the Mediterranean for decades—perhaps 80 years—before it finally sank.

It found a new home in a Crusader castle built about A.D. 1200 that overlooks Kyrenia. Conservators there reconstructed the ship (opposite) for display in a vaulted gallery of the castle.

marked the wreck. The distinctive carrot shape of the jars (above) dated the vessel to the fourth century B.C. Most of the more than 400 amphorae recovered from the site were loaded aboard at the island of Rhodes, then famous for its inexpensive wine, and contained that drink.

In the last week of the 1968 expedition, divers uncovered the pine-ribbed hull of the Kyrenia wreck. The bulk of the 40-foot skeleton lay buried under two feet of hard-packed silt.

Greek shipwrights worked in the reverse of modern tradition, in which ribs of a vessel are fitted first. They assembled the shell of the hull first. After it was complete, they attached the ribs. Copper nails held the planking together, and the entire exterior of the ship was sheathed in lead, fastened with copper tacks.

The team wanted to save the timbers and reassemble them into the form in which they were found. The hull was dismantled piece by piece and moved ashore. A coded plastic label was attached to

CHAPTER SIX

THE ROMANS

753 B.C. — A.D. 475

Fishermen cast nets and poles in a Roman mosaic from a funerary chamber of the Necropolis of Hermes.

MARITIME ROME

BY ROBERT D. BALLARD

THE MYTHS OF THE FOUNDING OF ROME AND THE RAPE OF the Sabine women attest that the Roman Empire began as a land-based power. To administer and supply their vast territory, however, the Romans became masters of the Mediterranean, sailing its waters from east to west and north to south. To find ancient shipwrecks, my methodology is to study the trade routes of ancient civilizations—where traffic was heavy and dangers were many. Based on historical accounts, one of the busiest ancient deepwater routes lay between the North African city of Carthage, near present-day Tunis, and Ostia, Rome's first major port.

Many sea routes converged in the Straits of Sicily, a narrow body of water connecting the basins of the eastern and western Mediterranean, and the Skerki Bank, just north of there in the central Mediterranean. Beginning in 1988, I was fortunate to be part of an interdisciplinary team on the Skerki Bank Project, the first comprehensive study of ancient wooden ships lost in the deep sea. The project benefited from highly sophisticated technology. The *NR-1,* a nuclear research submarine capable of diving to 3,000 feet, was equipped with powerful sonar, a specialized lighting system, and a manipulator arm capable of retrieving relics from the seafloor. Ships that sink in deep water tend to be swamped and sink intact, while those that founder in shallows often break up. Part of our job was to study the condition of ancient wooden ships lost in the deep sea—some 95 percent of the ocean—to determine the area's importance to marine archaeology.

Our survey area included not only sea routes between Africa and Rome but also those between the eastern and western Mediterranean. In four expeditions, our team discovered eight shipwrecks in waters up to 3,000 feet deep, some hundred miles north of Carthage—the largest concentration of ancient ships ever found in the deep sea. All probably sank in storms. The *NR-1* and the optical tow-sled *Argo* located numerous elongated lanes on the seafloor, strewn with amphorae, indications that the ancient mariners tried to lighten their loads by dumping some cargo overboard.

We decided to map, survey, and sample the oldest of the wrecks: five wooden trading vessels of the Roman Empire dating from 100 B.C. to A.D. 400. All were the deep-bellied, square-rigged ships typical of the Roman era, and their varied cargoes represented the wealth of the empire. Among them were wine, olive oil, and honey. Such cargoes were stored in terracotta amphorae, and archaeologists can date a wreck by the characteristic shapes of the containers. Probably the largest ship, about a hundred feet long, was the earliest, dating from the first half of the first century B.C. It carried ten different forms of amphorae from Italy, Gaul, North Africa, and Greece. A second, and the most heavily laden of all the merchant ships, was filled with high-quality, rough-cut granite blocks and columns ready for builders, and hundreds of carefully packed sets of clay pots and plates.

At a depth of 2,600 feet lay the remains of another vessel. We named it *Isis* after the goddess worshipped by pagan seafarers throughout the Mediterranean. *Jason* retrieved from it a small terra-cotta oil lamp that enabled us to date the wreck from the last quarter of the 4th century A.D. A 2003 expedition followed up on our discoveries and further mapped the area of the Skerki Bank Roman wreck sites and the seafloor topography beneath the trade route that connected the bank to the western tip of Sicily. Our research of that summer showed that major areas of the Mediterranean have been trawled extensively, so that any remaining ancient shipwreck sites probably lie in deep water where commercial fisheries do not exist.

That same summer we excavated artifacts from a well-preserved Roman wreck in the Black Sea. Each wreck we find is a time capsule. Like a huge museum, the deep sea preserves history; it is where we must go in the future to rediscover our maritime past. ■

"

EVERYONE OF THOSE SHIPS WAS
WRECKED. IN ONE DAY
NEPTUNE SWALLOWED UP 30,000,000
SESTERCES ON ME. D'YE THINK
I LOST HEART? NOT MUCH!
I TOOK NO NOTICE OF IT, BY
HERCULES! I GOT MORE SHIPS MADE,
LARGER, BETTER, AND LUCKIER ...

"

PETRONIUS ARBITER,
"THE BANQUET OF TRIMALCHIO"

POTTERY PRIZES

In the summer of 2003, Robert Ballard and his team used the remotely operated vehicle Hercules *to recover artifacts from a first century B.C. Roman wreck in the Skerki Delta at a depth of about 2,400 feet. With the ROV's sophisticated manipulator arm, the operator maintains a sense of touch. As* Hercules *hovers near the exposed cargo in the hull of the ship, its basket-like manipulator is extended (right), and its arm clamps a long-necked jug (above), which it places in an elevator for removal to the surface. The site contained a wide variety of*

A LAUREL WREATH, *symbol of victory, crowns an ivory bust of Julius Caesar. Most famous of Romans, he spurred the city's transition from disintegrating republic to mighty empire prior to his assassination in 44 B.C.*

ROAD TO EMPIRE

FROM HUMBLE BEGINNINGS, A RUDE COLLECTION of mud-and-stick huts on a hill south of the Tiber River, the Romans built the richest and strongest empire in Western history. The Romans called the Mediterranean *Mare Nostrum*, "our sea," and for several centuries before Christ until several centuries after his death, Rome grew to spread its civilization to all the shores of that sea and far beyond inland.

Until Alexander the Great created his Hellenistic Empire in the 4th century B.C, the Western world was a collection of city-states. And even Alexander's great empire could not survive his death; it rapidly fell apart. But the Romans were consummate empire builders. Their greatest talent was government, and for centuries one code of laws, one army, and one emperor unified a multiracial Roman world of some 50 million inhabitants that swept across three continents—through the heart of Europe, a large part of western Asia, and the northern coast of Africa, from the Atlantic Ocean to the Red Sea.

Rome took to the sea reluctantly. Its strength lay in its army marching across the land, but as military conquests added to empire, the Romans increased their maritime power. Eventually the acquisition, then the administration, of its far-flung territories transformed Rome into a master of the seas.

According to legend, Rome was founded on April 21, 753 B.C. by the wolf-raised human twins, Romulus and Remus. Archaeological finds confirm that the hills along the Tiber River were inhabited in the ninth century B.C. by farmers and herders of the Latin and Sabine tribes. Early Rome was an assemblage of individual communities on different hills around the Tiber. Some scholars postulated that an Etruscan empire stretched across Italy; others now theorize that this was not so much an empire as a sphere of influence in which Etruscan culture held sway.

At first Rome was ruled by kings (753-509 B.C.), the last of which are viewed as Etruscan. The Etruscans, greatly influenced by Greek culture, introduced to Rome heavily armed infantry units, which they had copied from the Greek citizen troops. The Etruscans were also great engineers; they drained the central area between the hills along the Tiber where the Roman Forum would later stand and transformed the collection of hillside villages into the city of Rome. Over time Etruscan and Roman culture and religion fused. The Romans adopted Etruscan knowledge, including a writing system based on the Greek alphabet, and the architectural and engineering skills that would prove essential in the growth of Rome. Three of the leading Roman gods—Jupiter, Juno, and Minerva—are derived from Greek gods through the Etruscans.

Around 509 B.C. the Romans deposed the last Etruscan king, Tarquin the Proud, and established a republican form of government. A small group of ambitious aristocratic families vied for public offices. Two consuls elected by an assembly served one-year terms. However, friction and competition between the elite patrician class and the plebeians, or commoners, would plague Rome for more than 200 years. Between 409-287 B.C., a period known as the Struggle of the Orders, strife ensued between the two classes, which led to a number of reforms. From these shaky beginnings, the Roman Republic grew to endure for almost 500 years.The Romans loved language

VESTIGES *of former glory, crumbling columns and broken walls of the Temple of Saturn rise from the Roman Forum, political and spiritual heart of the empire. According to legend, Romulus and Remus founded the Eternal City here in the eighth century B.C.; archaeologists began excavations in the 1800s.*

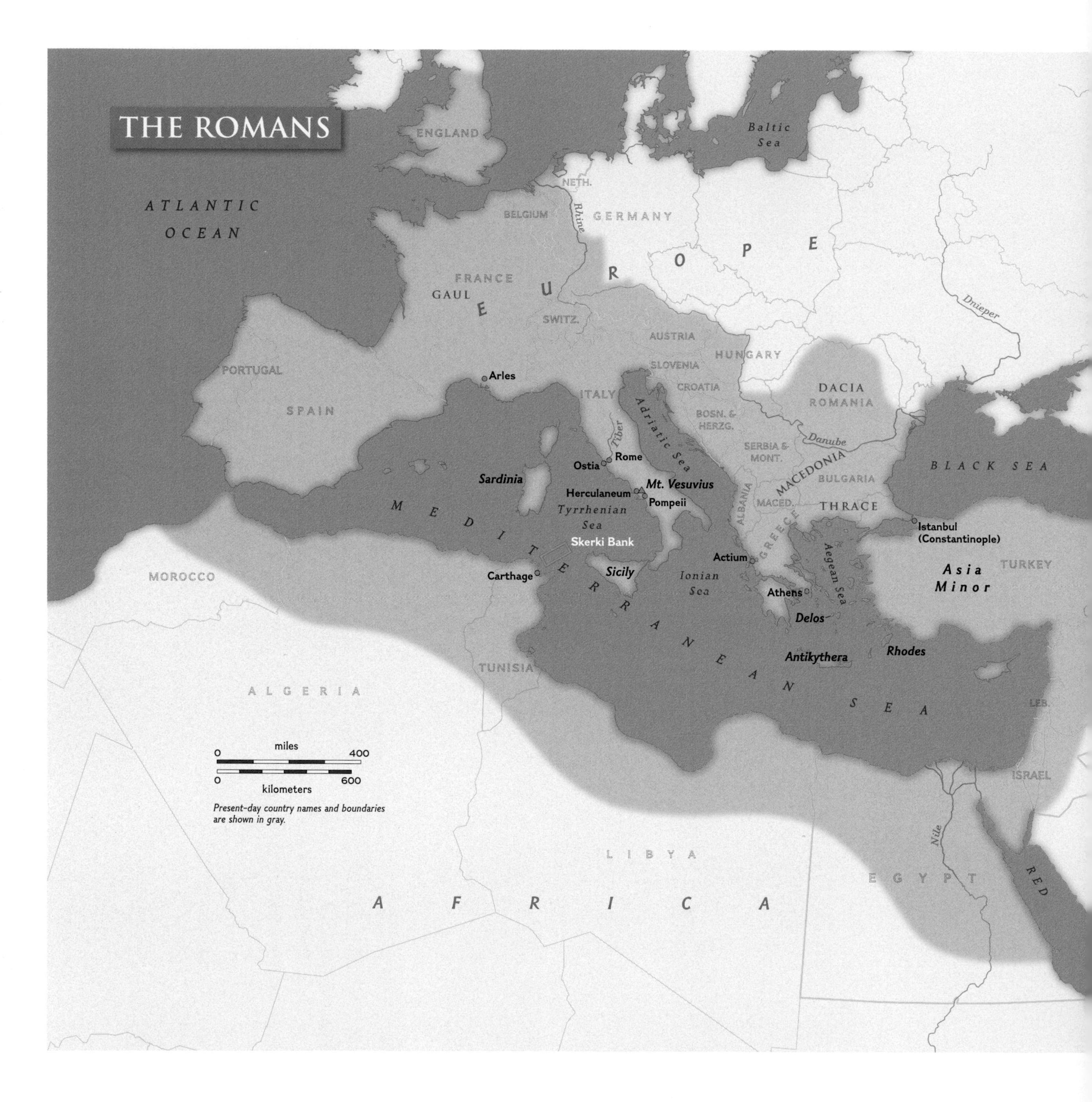

THE ROMANS
ATLANTIC OCEAN
ENGLAND
Baltic Sea
NETH.
BELGIUM
Rhine
GERMANY
E U R O P E
FRANCE
GAUL
SWITZ.
AUSTRIA
HUNGARY
SLOVENIA
CROATIA
PORTUGAL
SPAIN
Arles
ITALY
Adriatic Sea
BOSN. & HERZG.
DACIA
ROMANIA
Danube
Dnieper
SERBIA & MONT.
MACEDONIA
BULGARIA
BLACK SEA
Tiber
Rome
Ostia
Sardinia
Herculaneum
Mt. Vesuvius
Pompeii
ALBANIA
MACED.
THRACE
GREECE
Istanbul (Constantinople)
Tyrrhenian Sea
Skerki Bank
Actium
Aegean Sea
TURKEY
Asia Minor
MOROCCO
Carthage
Sicily
Ionian Sea
Athens
Delos
Antikythera
Rhodes
M E D I T E R R A N E A N S E A
TUNISIA
LEB.
ALGERIA
miles
0
400
0
600
kilometers
Present-day country names and boundaries are shown in gray.
ISRAEL
LIBYA
Nile
EGYPT
RED
A F R I C A

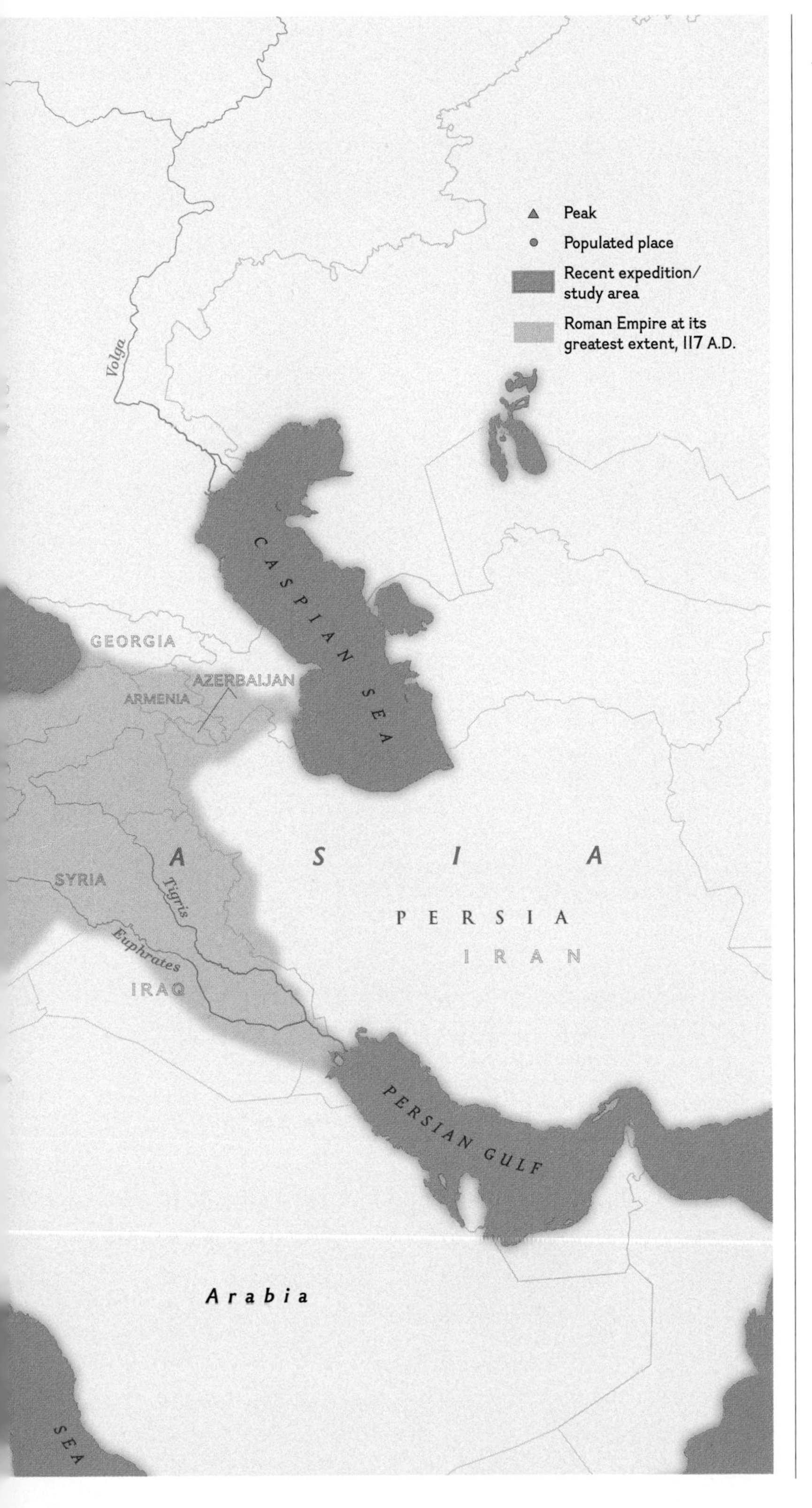

and enacted and recorded their laws. They left a wealth of written records, which aid modern researchers in reconstructing the Roman cultural achievement. Many scholars consider Rome's greatest legacy to be the comprehensive body of statutes and case law they established. Written laws to protect individuals from one another and from the power of the state was another idea they borrowed from the Greeks, and a concept they put into daily practice.

The Roman Republic began with skirmishes with nearby rival states, and one of its first accomplishments was a formalized treaty with its Latin neighbors. The Romans subjugated the Samnites to the east and defeated the Greeks to the south, and by 290 B.C., the Romans were masters in central Italy. From the start, one of the keys to the Romans' success was that they were flexible rulers. Administration at the local level could be open and tolerant. As the Romans consolidated their rule in Italy, they created a confederation, evolving a complex system of conveying status to their allies, including, at times, Roman citizenship. To its loyal citizens, Rome granted stability and financial success unequaled in the ancient world.

The growth and success of Rome was also partly due to the city's geographic setting. The city was located far enough from the sea to escape the pirates that raided the Mediterranean, and the site commanded the first practical crossing place over the Tiber, the major river of central Italy. The Tiber was the water road to the interior, with its agricultural and material resources. Rome could control traffic and goods from northern to southern Italy and trade along the river and the areas bordering it.

Rome's great military strength lay in its armed citizen infantry, a concept inherited from the Etruscans. The Romans produced a superbly trained, organized, and disciplined corps of foot soldiers. These large standing armies fought hand to hand, or with catapults that hurled heavy stones, and penetrated fortifications with movable scaling towers and battering rams.

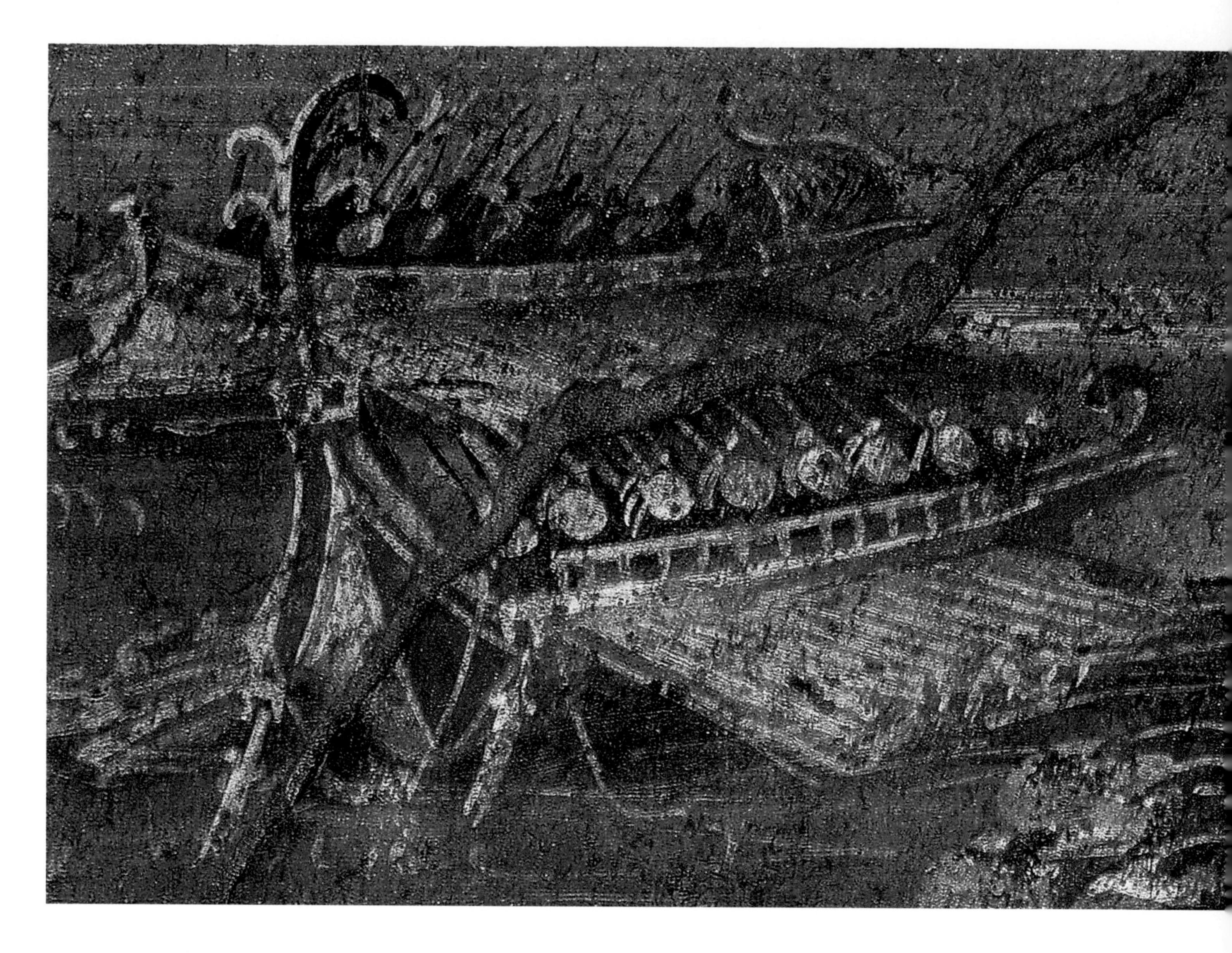

Despite its formidable legions, Rome could not compete at sea; it possessed no war fleet like that of Carthage or some of the Hellenistic city-states. Rome's first navy, established in 311 B.C. to protect coastal towns from pirates, consisted of 20 galleons. It had not a serviceable ship remaining by 263 B.C.

A minor incident in Sicily ignited the first of the Three Punic Wars against Carthage for dominion of the Mediterranean (264-146 B.C.). In 260 B.C., during the early stages of the First Punic War (264-241 B.C.), the Roman senate mandated the building of 120 ships in two months. Using a model of a Carthaginian ship that had run aground, the Romans produced slower but stronger vessels. Their new fleet—floating infantry carriers—contained an innovation. At the sea battle of Mylai that same year, the Romans deployed the new device called the *corvus* ("raven"), a gangplank with a large spike at its lower outboard side. The corvus, mounted in the bow, was essentially a boarding ladder attached with a swivel mount to the deck at one end and held upright against a mast. Tackle allowed for its movement. When a Roman galley came within range of an enemy ship, the corvus was dropped and the Romans stormed aboard, turning the sea battle into a virtual land battle. With newfound success, the Romans proceeded to augment their navy, although they continued to harbor an antipathy to seafaring.

ROMAN MARINES *crowd four ram-prowed warships in a wall painting from Pompeii. Although the Romans adopted the bronze-plated beak design favored by navies in the Mediterranean since the first millennium B.C., they preferred to grapple and board enemy vessels rather than ram them. Marines were equipped with shields, spears, and swords as well as grapples.*

By the end of the First Punic War, Rome had acquired Sicily, its first possession outside Italy. During the Second Punic War (218-201 B.C.), the Romans repelled an invasion by Carthaginian leader Hannibal from across the Alps and conquered much of Spain. In the Third Punic War (149-146 B.C.) the Roman navy blockaded Carthage, and Roman soldiers burned the city to the ground. Once Carthage was in flames, Rome faced little competition and took control of the western Mediterranean and part of North Africa.

After the Punic Wars, the Romans began to extend their empire eastward. The island of Rhodes, the most powerful of the small maritime states of the Aegean, took an aggressive stance against piracy. In 168 B.C., when the Rhodians did not support Rome against Macedonian King Perseus, the Romans destroyed Rhodes financially by declaring Delos a free port. The power of Rhodes faded; with no further Rhodian patrols to stop piracy, attacks became a problem.

Rome bested them, too. In 67 B.C. the Roman senate appointed a veteran soldier named Pompey to eliminate the raiders. He divided the Mediterranean into sectors, appointed commanders to attack simultaneously, and freed it and the Black Sea of attack within a few months.

In 60 B.C. three great military leaders, Pompey, Marcus Crassus, and Caesar, formed an alliance known as the First Triumvirate to rule the republic. By this time Rome had become a hotbed of political intrigue and civic unrest. Patricians and plebeians had struggled throughout the years of the republic, with the plebes gaining more political power. Social unrest had threatened to bring about the downfall of Rome.

FLOWERS *ring the remains of Carthage (opposite), razed by Rome in 146 B.C. A Punic stela (above), recalls its Phoenician heritage. "Punic" in Latin means Phoenician.*

Republican Rome consisted of three classes of people: the patricians, the peasantry, and slaves. The patricians, descendants of the first ruling families who had served in the senate, grew increasingly rich and decadent during the republican period and invested in new forms of self-indulgence and ostentation. The poor got poorer. The peasantry was conscripted into the Roman legions; many migrated to towns, where they tried to eke out an existence as shopkeepers or craftsmen. The peasant population base, which made up the essence of the Roman legions, shrank as the peasants migrated to the cities. Their land was bought up by members of the moneyed class, who employed slaves bought or captured to work the huge estates. The Romans viewed slavery as part of their social hierarchy. Some slaves, however, managed to buy their freedom or were manumitted for acts of service to their owners. Such freedmen made up a large part of the artisans and shopkeepers in Roman society. They lived a precarious existence, though, in crowded multistory tenements, sleeping on beds of straw, cooking on open braziers, and facing constantly rising prices for essentials such as grain. The rich, on the other hand, lived in luxury, in airy mansions surrounding open courtyards.

Meanwhile, republican Rome established the buildings that gave the city its enduring character. The ancient public square, the *Forum Romanum,* a civic center on low ground between the Capitoline and Palatine Hills, grew in size and importance, becoming the heart of the political, legal, commercial, and religious life of Rome.

In 58 B.C. Julius Caesar began the conquest of Gaul (modern France). By 55 B.C. he moved his troops farther north to Britain, but the senate ordered him to disband his army. Caesar, victorious and ambitious, defied the senate's order and returned to Rome in 49 B.C. to declare himself dictator. His rule brought an end to the Roman Republic, but his lack of deference to conservative opinion aroused jealousy and anger. Concerned about his increasingly autocratic actions, some nobles decided to assassinate Caesar. The conspirators who murdered him on the Ides of March in 44 B.C. sought to save Roman democracy, but instead they brought about a civil war that continued until Octavian—Caesar's grandnephew and heir—defeated

SAILS LOWERED, *ships enter a harbor in a fresco from Stabia, Italy (right). A third century A.D. plaque (top) evokes the man-made Roman port of Portus. A merchantman glides toward a gigantic statue of Neptune that stood outside the harbor; behind the ship, a flaming beacon crowns the lighthouse that guarded its entrance. A 16th-century painting reveals both landmarks (above) as well as the semicircular outer basin and the hexagonal inner basin of the port.*

TIMEWORN MOSAICS *advertise different businesses at the ancient port of Ostia on the Tiber (opposite). The port floundered as the river silted in but revived when Emperor Trajan completed a canal from Ostia to Portus. Ostia, home to many of Portus's workers, boasted some 50,000 inhabitants, who patronized shops like the poultry vendor's carved in a Roman funerary stele (above).*

Mark Antony and Egypt's Queen Cleopatra in the battle of Actium in 31 B.C. Octavian assumed the name Augustus, established a one-man rule in Rome, and became its first emperor in 27 B.C.

As a republic, Rome had conquered many lands, extending its sway from the Iberian Peninsula to Syria, and from North Africa to France. Thus when Augustus initiated his system of rule, known as the Principate, he ruled an empire.

The first century B.C. was what historians call the classical period of Roman history. Greece was the inspiration for much of Rome's blossoming culture. The deities of Greece were renamed and worshiped at Roman temples cast in Greek style. Classical Rome's religion was one of rituals rather than of faith or belief. The objective of the Romans was to propitiate their gods or inspire them to action with rituals, vows, or sacrifices. Cicero's impassioned speeches on the inherited values and institutions of Roman society inspired the Roman masses. Poets Horace, Virgil, and Ovid espoused the political and social roles of poetry and extolled the might and honor of Rome. Livy immortalized its history.

In architecture and engineering, as in warfare, the Romans excelled. Art followed Greek models, although while the Greeks emphasized proportion and line, the Romans highlighted power and exuberance. The Romans invented *caementicum*—cement, a mixture of volcanic ash, lime, water, and fragments of stone or rock added for strength and color. Theirs was the first concrete strong enough to hold bricks and stones together, and it enabled the Roman builders to develop the arch, and subsequently to erect vaults and domes. The Greek column, necessary for structural support, became a decorative element in Roman buildings.

Of all the triumphs of Roman engineering, their water works stand out. Between 312 B.C. and A.D. 226, they

constructed 11 aqueducts to bring water to a growing Rome. Some drew water from 50 miles away. Rome's famous fountains had a practical purpose. As water flowed downhill through an aqueduct, pressure built up in the line. Where the water came into the city, the Romans designed fountains to relieve that pressure. Rome's famous Trevi Fountain is the terminus of an ancient 13-mile-long water line. Engineers completed the planning, but Roman soldiers provided most of the labor for these waterworks. They also built forts, bridges, and roads.

When Augustus became Rome's emperor, his first real task was maintaining and administering his vast realm. He created a professional army of volunteers organized in legions—possibly the most efficient military machine in history. Some 30,000 legionnaires were stationed in permanent camps to guard the empire's frontiers. Roman culture spread with its armies. Settlements quickly took on the hallmarks of a Roman town—a protective wall, paved streets, sidewalks, temples, and a public bath—most built by legionnaires. Thousands of miles of highways knit the empire together, speeding Roman trade and influence. Initially, conquering Romans allowed local areas to govern themselves; later, Roman governors were assigned.

The rule of Augustus was essentially a monarchy, but he abided by the laws of the senate and voluntarily restricted his activities to those that would not offend the people of Rome. In the capital, dense, crowded, and dirty, Augustus focused on improving and rebuilding Rome, turning it from a city of brick to a city of marble. Augustus finished formulating the 12-month, 365-day calendar begun by Julius Caesar that we still use today.

The Battle of Actium had convinced Augustus of the vital importance of a navy to the maintenance of empire, and he reorganized and maintained a standing fleet. Imperial Rome became the greatest maritime power in the world. Augustus also created a solid administrative system, which allowed the empire to remain relatively stable even after his death in A.D. 14 under a number of successors of varying capabilities. The matter of succession always remained a problem, however.

Provincial affairs were administered adequately, even though politically ambitious aristocrats and emperors as despicable as Nero brought intrigue and scandals to the capital. Vespasian, emperor from A.D. 69 to 79, quelled a civil war, restoring order to Rome. He also initiated the construction of the Coliseum. It took more than a decade to complete the arena, with seating for the 50,000 urban citizens who enjoyed the spectacles designed to keep them satisfied with Roman rule. Trajan, A.D. 98-117, continued to glorify Rome, building a new forum and market. Under his leadership, the empire reached its greatest extent, encompassing 40 provinces—parts of more than 40 modern nations. Half the size of present-day China, the Roman Empire covered almost two million square miles and stretched from Europe to Africa and the Euphrates River. Romans enjoyed some 250 years of relative peace and order, which began with the rule of Augustus. The cities of the eastern empire recovered from battles; the Roman way of life was introduced in the western empire; and Rome grew in naval and commercial might.

Since the time of Augustus more than a hundred merchant ships a year sailed from the Roman port of Ostia to African and Indian ports. Trade fostered unprecedented opulence in Rome. Patricians built mansions of marble imported from Greece and North Africa, decorated them with furniture crafted from exotic woods imported from Asia Minor, feasted on Indian pheasants served on plates of Spanish gold, and dressed in silks and gems from the far reaches of the Orient.

Beyond the riches of the Mediterranean, where sailors battled rocky hazards and unpredictable weather, Roman commerce relied on two great trade routes. The first one—the overland route—crossed central Asia. The second one led from the Mediterranean up the Nile, overland to the

THE FOUNDING OF ROME: ROMULUS AND REMUS

When the Trojan War ended, the Greek hero Aeneas sailed to the country of the Latins, where he wed the daughter of a king and established a line of monarchs who lived in a city called Alba Longa. Many years later, a wicked descendant drove out his elder brother, whose daughter, Rhea Silvia, was forced to become a priestess to the goddess Vesta.

Rhea Silvia, dedicated as a Vestal Virgin, was forbidden, on pain of death, to love any man. Mars, the god of war, however, wooed and won her. When Rhea Silvia bore twin boys she named Romulus and Remus, the town punished her by bricking her—alive—in her own tomb. Her sons were to be drowned. When the servants assigned to dispatch the twins arrived at the Tiber, they found it overflowing its banks in full flood. They could not get close enough to dump the boys into the water; instead they set the babies' basket down on the shore, and the roiling waters soon carried it away. Romulus and Remus survived a wild ride over boulders and shallows and raging waves, until the basket snagged on tree roots and spun into a backwater shaded by a fig tree where animals came to drink.

When the twins looked around, they were gazing into the yellow eyes of a huge she-wolf. Suspended from her jagged teeth, first Romulus, then Remus, traveled to her lair and joined her litter. The she-wolf nursed the two boys as she did her own cubs, and they

A bronze cast in the sixth century B.C. recalls the wolf that nursed Romulus and Remus.

The Roman settlers select wives in The Rape of the Sabine Women, *an oil on copper painted by Friedrich Christoph Steinhammer in 1622.*

grew in size and strength. One day a herdsman hunting for wolves that had killed some of his livestock came upon the two boys. The herdsman knew that the law demanded their death. However, he loved his own sons so much that he took Romulus and Remus home and raised them as his own children in a thatched hut on the Palatine Hill.

Romulus and Remus grew into brave—but quarrelsome—boys. They were united in their determination to found a city to rival Carthage, Troy, or Athens, but they fought over its location. Romulus demanded the Palatine Hill as the center of the new town; Remus insisted on the Aventine Hill. When Romulus plowed a furrow around the Palatine to serve as a sacred boundary and threw the dirt up to build a wall around the site, Remus jeered and knocked the barrier down. Romulus grabbed a nearby boulder and bashed Remus in the head. He died instantly, falling to the ground bathed in blood. "So perish whoever else shall overleap my battlements," swore Romulus, and he built the city of his dreams.

Rome was populated solely by male settlers, and they were mostly runaways and criminals of central Italy who did not attract wives.

Romulus fretted about Rome's future and devised a plan. He invited the neighboring Sabine tribe to a festival to view the completed city. As the Sabines relaxed at the feast, the Romans seized all the Sabine women and carried them to the Palatine fortress. Over time, the Sabines grew content with their Roman husbands and raised families. When their tribe finally returned to rescue them, the women rushed between the enemy lines and demanded peace between the Sabines and the Romans.

Romulus expanded Roman rule and established many political institutions. Years later, during a storm, he was snatched up to heaven, where he became the god Quinius. "My Rome," he later prophesied in the dream of a Roman elder, "shall be capital of the world."

Red Sea, down the Red Sea, and across the Indian Ocean. In ancient times, navigation was an art rather than a science, with few instruments available beyond the sounding lead. The key to maneuvering the long sea route lay in the weather. From Italy, ships departed for Egypt in April and rode the trade winds south for two to three weeks. The return trip, against prevailing winds, might take a heavily laden vessel two months. The regularity of monsoons in the Indian Ocean and Arabian Sea determined the sailing season there. From May to September, ships rode monsoon winds from the southwest to the spice ports of India. For the remainder of the year, the winds blew from the northeast, speeding ships back to the Red Sea.

Roman merchant vessels were roomy and round, with a broad square sail. The normal-sized merchant ships carried 100 to150 tons burden, but larger ships of 350 to 500 tons were hardly a rarity. The huge Roman grain ships may have reached 1,300 tons burden. Like Athens, imperial Rome depended on imported grain to feed its growing population. During the years of the republic, politicians currying favor had offered free or subsidized grain to the Roman masses. By the second century A.D. more than 300,000 tons of grain from Sicily, Sardinia, Egypt, and North Africa—all of it exacted as taxes by Rome—made its way to its port at Ostia and then to the capital. Smaller craft could sail from Ostia up the Tiber River, but as larger ships became more economical to run, the Romans developed a system of river barges. Ships were unloaded in Ostia and their cargoes carried in these river craft from the coast to Rome.

Sediments flowing down from the Tiber River silted the harbor, and as commerce increased Ostia could not handle the numbers and sizes of Rome's commercial fleet. Ostia's limited facilities contributed to a grain shortage in Rome, and in A.D. 42 Emperor Claudius began construction of the largest man-made harbor in the ancient world: Portus. On the marshy coast two miles north of the mouth of the Tiber, two breakwaters almost a half-mile wide and 50 feet long were built to circle around a 160-acre anchorage. Three hundred ships could dock at Portus's outer basin, and a 78-acre inner basin was surrounded by wharves and warehouses. A mammoth three-story lighthouse guided merchant ships carrying cargoes from all over the known world. The goods then moved 15 miles upstream to the docks of the capital.

In A.D. 103 Emperor Trajan enlarged Portus and built a deep canal connecting it to Ostia. As Portus grew, Ostia also flourished. It housed many of the workers of Portus and most of its administrative and business agencies. Some 50,000 people lived in the bustling seaport, most of them freedmen or foreigners.

Large trading companies were owned by wealthy investors of many different nationalities. A growing number of merchants and businessmen became rich as trade expanded. In Ostia, Portus, and Rome workers in various jobs banded together in local trade groups and associations called *collegia.* Soon more than 150 corporations existed in Rome, each dealing with a specific type of cargo or profession. At the height of Rome's commercial empire, in the second century A.D., it was one of the richest and most powerful cities the world has ever seen. At its peak, Rome housed perhaps one million residents.

Roman tastes had also grown sophisticated. The most eagerly awaited ships were those bringing vital grain, but others carried the commodities Rome's craftsmen needed and the luxuries its wealthy demanded—amber from the Baltic, glass and textiles from Britain, lead from Spain, horses from Thrace, gold from Dacia (in modern Romania), the famous purple dye from Phoenician Tyre, linen from Egypt, and prunes, plums, and medicines such as sulfides and saffron from Asia Minor. From the Red Sea, then overland to the Mediterranean, came frankincense from Arabia and myrrh from Africa. The leopards, lions, and panthers that competed in the wildly popular public

EXERCISING *before a bath, Roman women lift weights in a fourth-century mosaic excavated on Sicily. Every sizable Roman town had at least one public bath. Patrons paid a nominal fee to enter, then worked up a sweat in the exercise yard, or* palestra, *before heading to a lukewarm pool followed by submersion in a steaming bath.*

LADIES OF LEISURE *converse, in a fresco unearthed in Herculaneum, a resort town destroyed by a two-day eruption of Mount Vesuvius in A.D. 79. Poisonous gases rapidly killed many of the town's 4,000 inhabitants. Murals, frescoes, and sculptures testify to the high standard of living Herculaneans enjoyed.*

TIERED SEATS *of two circular theaters accommodated the 10,000 inhabitants of the port and vacation retreat of Pompeii until ash and lava flows from Vesuvius, looming six miles north of the city, buried it. Excavations began in 1860 and still continue today.*

DELICATE GLASS *and finely crafted jewelry offer proof of the luxurious life led by Pompeiians—a standard of living not achieved again until the 19th century. The excavation of Pompeii marked the dawn of modern archaeology and offers the most detailed picture of a prosperous Roman town in the first century A.D.*

games at the Coliseum were imported from Africa and Asia Minor. The peacocks that were a delicacy enjoyed by the rich arrived from India. From the Indian Ocean route came ivory and spices, such as pepper, cloves, and cardamom from India, and even silks from China, some 8,000 miles distant.

Rome exported finished goods crafted by its many artisans—jewelry, cut gems, glassware, metalware, pottery, and wine, many produced from imported raw materials. Third-century A.D. accounts note that Roman ships visited what is probably now Hanoi, Vietnam, and reached as far as Malaysia, Java, and China. Gold and silver paid for most of the imported goods. Roman coins have been unearthed in Africa, Afghanistan, India, and Indochina.

Trade declined during the third century A.D.—and with it, so did the Roman empire. Pampered by the sybaritic luxuries marine commerce offered, Rome's citizens grew lazy and spoiled. They wanted bread and the circuses of the gladiatorial games, not work or conquest. The empire's magnificence began to fade. Imperial control weakened as the administration and defense of the empire grew increasingly unwieldy. From A.D. 161 to A.D. 180, Emperor Marcus Aurelius spent his rule fighting four wars and a plague that swept the capital; the long-standing peace of the Roman world began to crumble. In a 50-year period beginning about A.D. 235, Rome had more than 26 emperors; only one of them died a natural death. The army created emperors—and deposed them. The Roman assembly's loss of control led to excesses by some rulers; and public confidence in leadership waned. A growing population drift to cities severely reduced the empire's agricultural manpower and output. Prices slowly rose. Fewer taxes were collected, curtailing Rome's income. An empire increasingly expensive to administer returned less capital to Rome, while its resources were wasted on battles between rival rulers. Barbarian invasions further weakened the Empire, and the aggressive Persians threatened its eastern provinces.

By 284, when the army selected Diocletian as emperor, Rome faced invasions on all its frontiers. Determining that the empire was too large to be administered by one person, Diocletian divided it in two: a western and an eastern province, each with its own administrator. In his efforts to tighten the organization of the empire, Diocletian began the systematic persecution of Christians. Early Christians refused to acknowledge the state religion, which to the Romans was equivalent to being disloyal to the state. Thus, Christianity was considered subversive, and Christians at times suffered for it. Nero had initiated mass persecutions, blaming Christians for the fire that swept Rome during his reign. In A.D. 302, however, Diocletian made the persecution of Christians an imperial effort rather than the result of any local initiative. Chaos followed Diocletian's death. As many as seven men vied for power.

Following the civil wars of A.D. 306-324, Constantine the Great emerged victorious and reunified the Roman Empire. He was sympathetic to the Christians and called an end to their persecution in A.D. 313 with the Edict of Milan. In A.D. 324, following the end of civil strife, he also passed laws in favor of Christianity. The church grew in size and power.

During his reign Constantine restored order and financial stability to the Roman Empire. He created a second capital city in Byzantium, renaming it Constantinople, and set the stage for a new division of the empire. The new capital's political reach moved toward the Greek-speaking eastern provinces. As Constantine lay dying, in A.D. 337, he converted to Christianity, setting an example that spread among the leading citizens of the empire and outward to its provinces. Rome, already the seat of the Apostle Peter, the designated successor of Jesus, became Christianity's base. During the reign of Emperor Valens

CREATURES OF *the sea crowd a Roman mosaic from the first century B.C. (opposite). Under Trajan (A.D. 98-115), the Roman Empire reached its greatest extent; the emperor had the largest and last of the imperial forums built. Trajan's Column (above right) celebrates his conquest of Dacia. The frieze that winds around it (above left) commemorates specific events of his two campaigns.*

(364-378), Christianity became the official religion of the Roman Empire.

Succeeding emperors faced the growing problem of barbarian invasions. The late fourth and fifth centuries A.D. were a time when migrating peoples stressed the Roman Empire. Pressure from the Huns, who defeated the Ostrogoths in A.D. 372, set the Visigoths in motion. Increasing numbers of Germanic peoples strained across Roman frontiers, robbing and looting as they moved. Many settled in the empire, dividing the land with its rightful owners. The defense of empire came to depend on Germanic leaders and on the loyalty of their barbarian troops. In A.D. 379 a Spanish officer, Theodosius I, was declared emperor. He orchestrated a treaty with the Visigoths. At Theodosius's death, in A.D. 395, the Visigoths, under their leader Aleric, rose up once more. Roman legions, besieged by Picts and Saxons, left England in the beginning of the fifth century. In A.D. 410 the Visigoths, under Aleric, sacked Rome; in A.D. 455 it was the Vandals' turn. The last Roman emperor, Romulus Augustulus, was deposed in A.D. 476 by German chieftain Odoacer, who was declared king by his troops, and a Germanic kingdom was established in Italy. The Western Roman Empire had run its course. The capital, at its height in the second century A.D., home to as many as a million people, decayed. Its population shrank to some

EXOTIC IMPORTS: *African animals came to Rome from northern Africa to serve in the circuses and gladiatorial contests at the Coliseum. In a detail from a third-century A.D. mosaic, handlers prod an elephant up the gangway of a trading vessel.*

17,000, and Rome reverted to its agrarian roots.

Despite the glory and grandeur commerce brought to Rome and its empire, the Roman world remained largely agrarian throughout its history. The majority of the inhabitants were engaged in agriculture, to raise food for the masses. Food was produced and consumed locally throughout much of the empire; the huge grain shipments from northern Africa and Egypt were necessary supplements destined largely for Rome and other cities. Farmers comprised about 90 percent of the empire's population. Their lives, unlike those of the rich revelers of Rome, were grim, poor, and short. The luxuries that contributed to Rome's decadence did not find their way far beyond the city, where manual labor was scorned.

Even during the late empire, the time of greatest opportunity to reach high office from a relatively humble background, there was little prospect for advancement for the great masses of the population. They lived just above, or close to, subsistence level. In addition, all the jousting for power and position in the Roman world involved men; it was a male-dominated society in which women could neither vote nor hold office. There is small mention of them and little is known of their roles outside that of wives and housekeepers.

Little effort was made to preserve the ancient treasures in Rome over the centuries, and many structures were cannibalized for their building materials. Two cities buried in a two-day eruption of Mount Vesuvius in A.D. 79 offer the most vivid clues to everyday life of the wealthy. Pompeii, an old port and popular vacation retreat for Rome's elite on the Bay of Naples, and Herculaneum, a

THREE LEVELS *of arches span the Gard River more than 160 feet above the water, near Nimes, France, in one of the best-preserved and most spectacular of Roman aqueducts still standing. Water ran through a conduit at the top of the structure, built in the time of Augustus to supply water to the Roman city of Nemausus.*

ROME'S MOST WESTERLY *African outpost, Volubilis, in present-day Morocco, boasted sumptuous aristocratic mansions floored with intricate mosaics (above) from the first century B.C. A 45,000-seat amphitheater (opposite), best-preserved in Africa and among the largest in the Roman world, survives in El Djem, Tunisia, the Roman settlement of Thysdrus.*

resort town on the western flank of the volcano, contain a legacy of Roman provincial life that archaeologists are still excavating. Work in Pompeii, six miles south of Vesuvius, began in 1860. The remains of some 2,000 people struck down and buried by the eruption have been found. The town of more than 10,000 residents and many hundred more vacationers was buried slowly beneath 12 feet of ash and rock. Many people may have had time to flee. At Herculaneum, surges of hot gases and pumice flooded the town quickly, killing and burying much of its population. Colorful murals, portraits, furniture, and Greek-inspired sculptures attest to the high standard of living.

Archaeologists have also long sought clues to Roman culture underwater. The first Roman cargo ship was found in 1864 at a building site in Marseilles, France. Greek sponge divers in 1900 found a Roman seagoing vessel off the island of Antikythera. It sank during the first century B.C., carrying tons of marble and bronze statuary. After the invention of the aqualung in the 1940s, a number of Roman shipwrecks were discovered in shallow waters of the Mediterranean. Until recently the study of shipwrecks has been restricted to the depths of scuba equipment—less than 200 feet. In 1989 and again in 1997, Robert Ballard and a team of scientists using highly sophisticated

technology located the largest concentration of ancient shipwrecks yet found in deep water. Four Roman ships dating to the late Republic and the early Empire periods lay at a depth of more than 3,000 feet on an ancient trade route between Ostia and Carthage. A sampling of amphorae revealed that they carried the empire's wealth, filled with honey, olive oil, and wine. Further research will reveal more information about their cargoes and their culture.

Roman ruins from Scotland to Germany, Spain to Iraq, and all around the Mediterranean attest to the vast holdings that made Rome in its heyday the center of the ancient world. In England the wall built by Emperor Hadrian stretched 73 miles from coast to coast, blocking northern invaders. In Arles, France, the capital of the province of Gaul, crowds still enjoy watching performances at an amphitheater built by conquering Romans. Remnants of the greatest highway system the world has ever known—53,000 miles of intercontinental roads—reveal three levels of substructure that the Romans built beneath the pavement.

From its capital in Constantinople, the eastern part of the empire persisted a thousand years after Rome fell. Finally, in 1453 Ottoman Turks conquered Constantinople, ending the Eastern Roman Empire as well.

Many scholars consider the legal code as Roman civilization's greatest achievement. Digests, institutes, and a revised legal code completed by Emperor Justinian in A.D. 534 form the foundation of Western law. America's founding fathers took Republican Rome as their model. As the empire in the West sank, the Christian Church preserved much of Rome's legacy. That heritage endures throughout the Western world in our language, literature, law, government, architecture, engineering, arts, and sports. ■

ANCIENT BARRIER: *Hadrian's Wall winds across desolate moorland near Housesteads in northern England. Farthest guardian of Roman rule, the 73-mile-long stone and turf wall protected Roman settlements in Britain from raiding northern tribes.*

HIGH-TECH DIVING FOR ROMAN SHIPWRECKS

During four expeditions beginning in 1988, Robert Ballard and his team on the Skerki Bank Deep Sea Project discovered the largest concentration of ancient shipwrecks ever found in deep water—five Roman vessels dating from about 100 B.C. to A.D. 400. All were found within a 55-square-mile area off western Sicily, along a direct-line sea route between Rome and the ancient North African settlement of Carthage, then a Roman province.

Marine archaeologists had been restricted to depths of about 200 feet, the limit for scuba divers. High-tech gear enabled Ballard and his crew to dive as deep as 2,800 feet. The U.S. Navy nuclear research submarine *NR-1,* equipped with powerful, long-range sonar, could find even small objects in wide areas of the sea bottom. After the members of the team located a target, they deployed the ROV *Jason.* Loaded with advanced hardware and able to grasp delicate glass objects with its netted hand and return them to the surface, *Jason* also had still and video cameras, sonar, and precision mapping ability to perform analytical work on the wrecks.

In the control room of the project's support ship, *Carolyn Chouest* (above), Ballard and Dana Yoerger, at left, the project's head of engineering, confer as they study videos telecast from the 1.5-ton ROV hovering half a mile below the surface of the sea.

The team raised only sample artifacts, leaving much of the cargo intact for further investigation. Testimony to the open-sea trade that flourished between Carthage and Rome, the amphorae settled in the silt (opposite) when the ships foundered. Such pottery jars carried wine, olive oil, and fish sauce. The wooden vessels, square-rigged and deep-bellied, probably sank in squalls. Their skippers, unlike the mariners who hugged the coasts, were risk-takers who braved the storms and dangers of deep-sea sailing.

In the nine-year period from 1988 to 1997, Ballard and his team recovered some 150 artifacts from the sites. All were untouched by looters, fishermen, or storms because of their great depth.

In 1988 Ballard and his team located the first of the Roman ships and named it *Isis,* after the goddess worshiped by many ancient Mediterranean seafarers. In 1989 an expedition returned, and *Jason* recovered 48 artifacts from the *Isis* wreck.

Medea, a metal platform fitted with lights and monitor cameras that served as a relay vehicle for *Jason,* was carefully lowered from the *Carolyn Chouest.* The *Medea* provided a steady base from which *Jason* maneuvered. In 1989 (above), *Jason* used its manipulator arm to lift an iron anchor from the *Isis* wreck 2,600 feet below the surface of the Mediterranean; a number of amphorae rest nearby. The tether that connects the ROV to *Medea,* which took the photograph, is visible at the very top of the picture.

The team also found amphorae, probably jettisoned during storms, scattered along the seabed. The containers, of various sizes and shapes, helped to determine the ages of the Roman vessels. The *Isis* proved to be the most recent, from the last quarter of the fourth century A.D. In 1997, using the NR-1, the team located the oldest of the wrecks, which dated to the first half of the first century B.C.

On the deck of the *Carolyn Chouest,* archaeologist John Oleson from the University of Victoria studies a large terracotta amphora used to store wine (opposite).

J Oleson

A NEW ERA OF DISCOVERY

BY ROBERT D. BALLARD

IN MORE THAN FIVE DECADES OF EXPLORING THE OCEAN and searching for history below the waves, I have been part of over a hundred expeditions—and I am always impatiently awaiting the next one. Before I was ten, the ocean had already captured my imagination and my devotion. Then I was content to watch crabs and sea anemones through glass bowls borrowed from my mother's kitchen. Since then I have been intent on finding new ways to delve into the depths.

Until recent years, marine archaeologists have been limited to a depth of 200 feet, the safe limit for scuba divers, leaving 91 percent of the ocean unexplored. Technological advances have afforded underwater archaeologists tools that open up that underwater wilderness. In expeditions to the Black Sea and the Mediterranean in the summer of 2003, Internet2 and a high bandwidth satellite link made it possible for scientists to work on the ocean floor from the comfort of their university laboratory. This real-time networking capability coupled with two underwater vehicles outfitted with high definition video cameras and a two-way audio link to shore greatly increased the number of scientists able to participate in the research. In the future a survey team aboard ship would consist primarily of engineers and technicians operating the exploration technology on a continuous 24-hour schedule with only a few scientists on the expedition. Once a discovery is made, the networking technology would be used to instantly network the scientists aboard ship via Internet2 to evaluate the significance of the discovery and recommend follow-up studies.

EYES WIDE, *Ballard marvels at images on the plasma screen TV in the control center aboard the* Knorr; *after more than eight hours of watching work on a Roman wreck in the Black Sea, he remained glued to the monitor. Following pages: From the bow, Ballard revels in day's end as the* Knorr *steams toward a new destination—and new discoveries.*

The *Hercules* vehicle system proved that wrecks could be scientifically and systematically excavated with minimum effort and maximum efficiency. Future work in the Mediterranean and in the anoxic bottom waters of the Black Sea are my first objectives for the near future. I know much more is down there. I'm going back—a lot. With technological advances and new opportunities for exploration come new responsibilities.

When I found the *Titanic* in 1985, I deliberately took little from the site. It is a memorial to those who died there, and I returned later and left a plaque in their memory. Other expeditions salvaged the vessel. The wrecks in the depths of the sea are museums; their relics add chapters to our knowledge. Yet the wrecks should remain largely untouched. Each one is a legacy that should be preserved for future generations. ■

KNORR

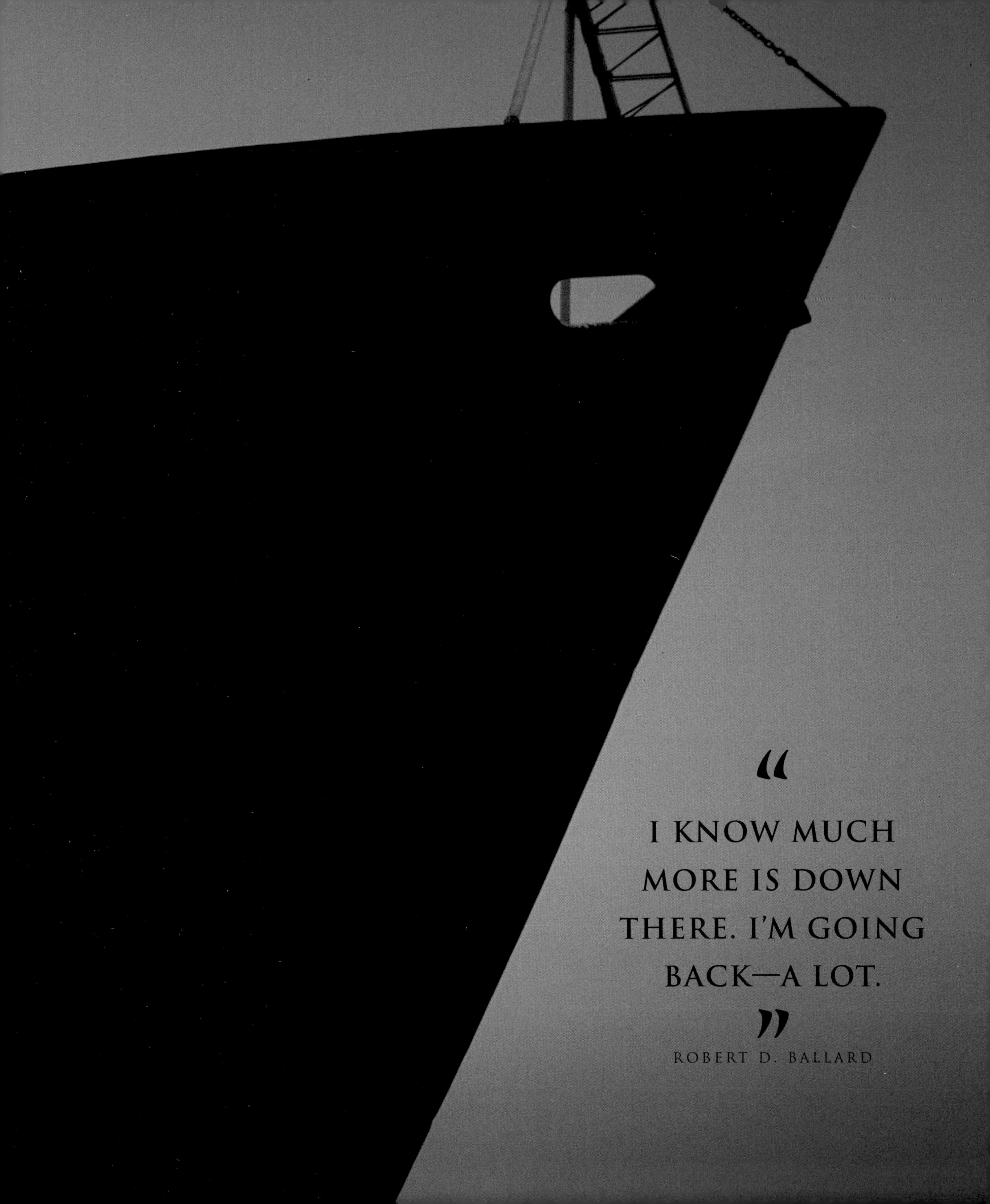

"

I KNOW MUCH
MORE IS DOWN
THERE. I'M GOING
BACK—A LOT.

"

ROBERT D. BALLARD

BIBLIOGRAPHY

Adkins, Lesley, and Roy A. Adkins, *Handbook to Life in Ancient Rome,* Oxford University Press, 1984

Ascherson, Neal, *Black Sea,* Hill and Wang, 1995

Baker, Rosalie F. and Charles F. Baker, III, *Ancient Greeks: Creating the Classical Tradition,* Oxford University Press, 1997

Ballard, Robert D., *Adventures in Ocean Exploration,* National Geographic Society, 2001

Ballard, Robert D. "The Discovery of Ancient History in the Deep Sea Using Advanced Deep Submergence Technology," *Deep-Sea Research* Part 1, Vol. 47, No. 9, pp. 1591-1620, Pergamon, 2000

Ballard, Robert D., et al. "Deepwater Archaeology of the Black Sea: The 2000 Season at Sinop, Turkey," *American Journal of Archaeology,* Vol. 105, No. 4, 2001, Archaeological Institute of America, 2001

Ballard, Robert D., et al. "Iron Age Shipwrecks in Deep Water off Ashkelon, Israel," *American Journal of Archaeology,* Vol. 106, No. 2, 2002, Archaeological Institute of America, 2002

Bascom, Willard, *Deep Water, Ancient Ships,* Doubleday, 1976

Bass, George F., ed., *A History of Seafaring Based on Underwater Archaeology,* Thames and Hudson Ltd., London, 1972

Bass, George F., "Beneath the Wine Dark Sea, Nautical Archaeology and the Phoenicians of the Odyssey," in *Occasional Publications of the Department of Near Eastern Studies and the Program of Jewish Studies,* Cornell University, 1997

Bass, George F., "A Prolegomena to a Study of Maritime Traffic in Raw Materials to the Aegean During the Fourteenth and Thirteenth Centuries B.C," Philip P. Betancourt and Robert Laffineur, eds., *TEXNH: Craftsmen, Craftswomen, and Craftsmanship in the Aegean Bronze Age,* Proceedings of the 6th International Aegean Conference, Philadelphia, Temple University, 18-21 April 1996, Aegaeum 16 (Liège and Austin 1997): 153-170.

Bass, George F., "Golden Age Treasures," National Geographic Magazine, March 2002

Boardman, John, et al. eds., *Cambridge Ancient History,* Cambridge University Press, 1982

Bunson, Margaret, *The Encyclopedia of Ancient Egypt,* Facts on File, 1991

Carcopino, Jerome, *Daily Life in Ancient Rome,* Yale University Press, 1940

Casson, Lionel, *Ships and Seafaring in Ancient Times,* University of Texas Press, 1984

Casson, Lionel, *Ancient Trade and Society,* Wayne State University Press, 1984

Clayton, Peter, and Martin Price, eds., *The Seven Wonders of the Ancient World,* Dorset, NY, 1988

Cotterell, Arthur, ed., *Penguin Encyclopedia of Ancient Civilizations,* Penguin, 1988

Cotterell, Arthur, ed., *Penguin Encyclopedia of Classical Civilizations,* Penguin, 1993

David, Rosalie, *Discovering Ancient Egypt,* Facts on File, 1993

Delgado, James P., ed., *Encyclopedia of Underwater and Marine Archaeology,* Yale University Press, 1997

Edey, Maitland, et al. eds., *The Emergence of Man—Lost World of the Aegean,* Time-Life Books, 1975

Empereur, Jean-Yves, *Alexandria Rediscovered,* George Brazilller Publisher, New York, 1998

Freeman, Charles, *The Legacy of Ancient Egypt,* Facts on File, 1997

Friedrich, Walter L., *Fire in the Sea, The Santorino Volcano: Natural History and the Legend of Atlantis,* Cambridge University Press, 2000

Gore, Rick, "Ashkelon, Ancient City of the Sea," National Geographic Magazine, January 2001

Hamilton, Edith, *Mythology,* Little, Brown, 1947

Johnson, Paul, *The Civilization of Ancient Egypt,* HarperCollins, 1978

Judge, Joe, "Minoans and Mycenaeans, Greece's Brilliant Bronze Age," National Geographic Magazine, February 1978

Katzev, Michael, "Resurrecting the Oldest Known Greek Ship," National Geographic Magazine, June 1970

Levy, Thomas E., ed., *The Archaeology of Society in the Holy Land,* Facts on File, 1995

Marinatos, Spyridon, "Thera: Key to the Riddle of Minos," National Geographic Magazine, May 1972

McCann, A.M., and J. Freed, "Deepwater Archaeology," Journal of Roman Archaeology Supplement, Ser. 13, Vol. 128

McCaughrean, Geraldine, *Roman Myths,* Simon & Schuster, 1999

Nardo, Don, *Life in Ancient Rome,* Lucent Books, 1997

Pomerov, Sarah B., et al. *Ancient Greece: A Political, Social, and Cultural History,* Oxford University Press, 1999

Reid, T. R., "The Power and the Glory of the Roman Empire," National Geographic Magazine, July 1997

Reid, T. R., "The World According to Rome," National Geographic Magazine, August 1997

Samage, Edwin S., ed., *Atlantis: Fact or Fiction,* Indiana University Press, 1978

Sigurdsson, Haraldur, *Melting the Earth, The History of Ideas on Volcanic Eruptions,* Oxford University Press, 1999

Starr, Chester G., *The Ancient Romans,* University of Michigan, 1971

Thubron, Colin, *Seafarers—The Ancient Mariners,* Time-Life Books, 1981

Wachsmann, Shelley, *Seagoing Ships & Seamanship in the Bronze Age Levant,* Texas A&M University Press, 1998

Wells, Spencer, *The Journey of Man,* Princeton University Press, 2002

CONSULTANTS

Alexander Bauer is an archaeologist specializing in the ancient maritime cultures of the Black Sea region. His Ph.D. is in anthropology from the University of Pennsylvania. He has been a member of the university's Black Sea Trade Project since 1997, where he is currently assistant director of the terrestrial survey team.

Lawrence E. Stager is the Dorot Professor of the Archaeology of Israel in the Departments of Near Eastern Languages and Civilizations and of Anthropology at Harvard University. He is Director of the Harvard Semitic Museum. For the past three decades he has directed terrestrial excavations in Cyprus (Idalion), Tunisia (Carthage), and Israel (Ashkelon). He was head of the archaeological team that surveyed two Phoenician shipwrecks, the oldest yet discovered in the deep sea. His work and writings have focused mainly on Bronze and Iron Age peoples of the Levant: Canaanites, Philistines, Israelites, and Phoenicians. His most recent book (co-authored with Philip King) is entitled *Life in Biblical Israel.*

Shelley Wachsmann is the Meadows Associate Professor of Biblical Archaeology at Texas A&M University's Nautical Archaeology Program. He has carried out extensive field work on shipwrecks in Israel and is the author of four books and numerous scientific and popular articles.

Special thanks to:

George F. Bass, professor emeritus of Nautical Archaeology at the Institute of Nautical Archaeology (INA) Texas A&M University.

Frederik Hiebert, professor of Near Eastern Archaeology at the University of Pennsylvania.

ACKNOWLEDGMENTS

This book is the product of so many great people who have worked with me over the years to help make the newly emerging field of Archaeological Oceanography possible as we have tracked down the ancient mariners of the deep. It has involved a wonderful mixture of scientists from many fields of research, including oceanographers like Dwight F. Coleman; engineering scientists like Dana Yoerger, Hanumant Singh, and Louis Whitcomb; historians like David Mindell and Brendan Foley; and archaeologists like Lawrence Stager, Fred Hiebert, Kevin McBride, Cheryl Ward, Anna M. McCann, also professor Francesco Torre, John Oleson, Jonathan Adams, Jennifer S. Smith, Kathryn Willis, Katherine Croff, Candace Major, Daniel Master, and so many of their colleagues and graduate students. It has also involved an incredible team of engineers and technicians who have developed and operated the magical vehicles we have used over the years to locate, document, excavate, recover, and conserve the remnants of the ancient mariners' shipwrecks beneath the sea. These great teams include Jim Newman, Cathy Offinger, Todd Gregory, Dave Wright, Sarah Webster, Mark Derose, Dennis Piechota, Cathy Giangrande, Martin Bowen, Andy Bowen, John Howland, and Jay Minkin. And the true unsung heroes of all expeditions we have made whom I wish also to thank are the officers and crews of the ships we have used including the *Northern Horizon* and the *Knorr* as well as the nuclear research submarine *NR-1.*

No cruise can take place without the people and organizations that sponsor the development of our technology and then sponsor the cruises to use the technology, which include the Office of Naval Research headed by Admiral Jay Cohen, NOAA's Office of Ocean Exploration led by Captain Craig McLean, and the Expedition Council of the National Geographic Society led by Terry Garcia. Helping me bring this all together have been Sandra Witten, Laurie Bradt, Andy Fedensky, and Trish Masse, who work behind the scenes each and every day.

INDEX

Boldface indicates illustrations.

CREDITS

1, Art Resource, NY; 2-3, Kevin Fleming; 4, Courtney Platt; 12-3, Art Resource, NY; 18, Robert Clark; 18-9, NG Maps; 20, Gianni DAgli Orti/CORBIS; 22-3, National Gallery Collection., By kind permission of the Trustees of the National Gallery, London/CORBIS; 24-5, NG Maps; 26-7, Robert Clark; 29, Ricki Rosen/CORBIS SABA; 30-1, The Art Archive/Musee Carnavalet Paris/Dagli Orti; 31, Gianni Dagli Orti/CORBIS; 32, David Lee/CORBIS; 33 (both), The Bridgeman Art Library; 35, Geoffrey Clements/CORBIS; 36, National Gallery Collection., By the kind permission of the Trustees of the National Gallery, London/CORBIS; 38-9, Ed Kashi/CORBIS; 40, Roger Wood/CORBIS; 41, The Bridgeman Art Library; 42 (le), Institute of Nautical Archaeology; 42 (rt), Gianni Dagli Ortii/CORBIS; 44, Archivo Iconografico, S.A./CORBIS; 45, Institute of Nautical Archaeology; 46-7, Courtney Platt; 48 (both), Bill Curtsinger; 48-9, Ned Seidler; 50-1, Gian Berto Vanni/CORBIS; 56-5 (all), Institute for Exploration (IFE) and the Institute for Archaeological Oceanography (IAO) at the University of Rhode Island Graduate School of Oceanography (URI-FSO); 58, Randy Olson; 62-3 (both), Marie-Louise Brimberg; 65 (both), Archivo Iconografico, S.A./CORBIS; 66-7, Priit J. Vesilind; 68, The Art Archive/Archaeological Museum Sofia/Dagli Orti; 69, Giraudon/The Bridgeman Art Library; 71, Archivo Iconografico, S.A./CORBIS; 72, James M. Gurney; 74, Paul Almasy/CORBIS; 76, Photographer/CORBIS; 77, Dave Bartruff/CORBIS; 78-9, Randy Olson; 79, Randy Olson; 80 (up), Photo 2003 National Geographic;80 (lo), Martin Bowen/WHOI; 81, Petar Petrov/Impact Press Group; 82-3, Richard T. Nowitz/CORBIS; 88-9, Erich LessingArt Resource, NY; 89, Institute for Exploration (IFE) and the Institute for Archaeological Oceanography (IAO) at the University of Rhode Island Graduate School of Oceanography (URI-FSO); 90, Art Resource, NY; 92-3, Kenneth Garrett; 96, Courtesy, Museum of Fine Arts, Boston. Reproduced with permission. c2000 Museum of Fine Arts, Boston. All Rights Reserved; 97 (up), K.M . Westerman/CORBIS; 97 (lo), The Bridgeman Art Library; 98, Kenneth Garrett; 98-9, Msssimo Listri/CORBIS; 100-1, CORBIS; 102, The Bridgeman Art Library; 102-3, Victor R. Boswell, Jr.; 104, Kenneth Garrett; 104-5, Art Resource, NY; 106-7, Tibor Bognar/CORBIS; 108-9 (both), Lee Boltin; 111, The Bridgeman Art Library; 112, Stuart Westmoreland/CORBIS; 113, Art Resource, NY; 114-5, Kenneth Garrett; 117, Guido Alberto Rossi/gettyimages; 118-9, Kenneth Garrett; 120-1, Stephane Compoint; 121, The Bridgeman Art Library; 122-3, Stephane Compoint; 124-5 (all), Stephane Compoint; 126-7, Joel Rogers/CORBIS; 132-3, Institute for Exploration (IFE) and the Institute for Archaeological Oceanography (IAO) at the University of Rhode Island Graduate School of Oceanography (URI-FSO); 134, The Bridgeman Art Library; 136-7, The Art Archive/National Archaeological Museum Athens/Dagli Orti; 140-1, Gianni Dagli Orti/CORBIS; 142, The Art Archive/National Archaeological Museum Athens/Dagli Orti; 143, Gianni Dagli Orti/CORBIS; 144, The Bridgeman Art Library; 145, Roger Wood/CORBIS; 147, Ruggerd Vanni/CORBIS; 148, Gianni Dagli Orti/CORBIS; 150, The Bridgeman Art Library; 151, Farrell Grehan/CORBIS; 152-3, Wolfgang Kaehler/CORBIS; 153, Kevin Schafer/CORBIS; 155 (LE), Roger Wood/CORBIS; 155 (RT), The Bridgeman Art Library; 156, Lauros/Giraudon/The Bridgeman Art Library; 157, The Bridgeman Art Library; 158-9, Stephen Johnson/gettyimages; 160-1, Brand X Pictures/gettyimages; 166, Institute for Exploration (IFE) and the Institute for Archaeological Oceanography (IAO) at the University of Rhode Island Graduate School of Oceanography (URI-FSO); 167 (up), Institute for Exploration (IFE) and the Institute for Archaeological Oceanography (IAO) at the University of Rhode Island Graduate School of Oceanography (URI-FSO); 167 (lo), Institute for Exploration (IFE) and the Institute for Archaeological Oceanography (IAO) at the University of Rhode Island Graduate School of Oceanography (URI-FSO); 168, Art Resource, NY; 170-1, Gianna Dagli Orti/CORBIS; 174 (up), The Bridgeman Art Library; 174 (lo), The Bridgeman Art Library; 174-5, Archivo Iconografico, SA/CORBIS; 176 (LE), Bettmann/CORBIS; 176 (CTR), Araldo de Luca/CORBIS; 176 (RT), Bettmann/CORBIS; 177 (LE), Gianni Dagli Orti/CORBIS; 177 (RT), Gianni Dagli Orti/CORBIS; 177 (CTR), The Bridgeman Art Library; 178-9, Erich Lessing/Art Resource, NY; 181, The Bridgeman Art Library; 182, ; 184-5, Yann Arthus-BertrandCORBIS; 187 (LE), Ruggero Vanni/CORBIS; 187 (RT), The Art Archive/Agora Museum Athens/Dagli Orti; 188, Scala/Art Resource, NY; 188-9, Metropolitan Museum of Art; 190, Bettmann/CORBIS; 192, John Heseltine/CORBIS; 193, Gianni Dagli Orti/CORBIS; 194-5, Archivo Iconografico, SA/CORBIS; 196-7, Bates Littlehales; 197, Bates Littlehales; 198, James Davis/Eye Ubiquitous/CORBIS; 199, Jonathan Blair/CORBIS; 200-1, Gilles MermetArt Resource, NY; 206-7, (both) Institute for Exploration (IFE) and the Institute for Archaeological Oceanography (IAO) at the University of Rhode Island's Graduate School of Oceanography (URI-FSO); 208, Bettmann/CORBIS; 210-1, Massimo Mastrorillo/CORBIS; 214-5, Art Resource, NY; 216, Roger Wood/CORBIS; 217, Nik Wheeler/CORBIS; 218 (up), Art Resource, NY; 218 (lo), The Bridgeman Art Library; 219, Art Resource, NY; 220, Erich Lessing/Art Resource, NY; 221, Araldo De Luca/CORBIS; 223, The Bridgeman Art Library; 224, The Bridgeman Art Library; 226, The Bridgeman Art Library; 227, Mimmo Jodice/CORBIS; 228-9, Roger Ressmeyer/CORBIS; 230, Mimmo Jodice/CORBIS; 232 (LE), Archivo Iconografico, S.A./CORBIS; 232 (RT), Vittoriano Rastelli/CORBIS; 233, Mimmo Jodice/CORBIS; 234-5, Art Resource, NY; 236-7, John & Dallas Heaton/CORBIS; 238, Gerard Degeorge/CORBIS; 239, Jonathan Blair/CORBIS; 240-1, Adam Woolfitt/CORBIS; 242, Priit J. Vesilind; 243, WHOI; 244, ; 245, Priit J. Vesilind; 247, Institute for Exploration (IFE) and the Institute for Archaeological Oceanography (IAO) at the University of Rhode Island Graduate School of Oceanography (URI-FSO); 250-1, Institute for Exploration (IFE) and the Institute for Archaeological Oceanography (IAO) at the University of Rhode Island Graduate School of Oceanography (URI-FSO)

One of the world's largest nonprofit scientific and educational organizations, the National Geographic Society was founded in 1888 "for the increase and diffusion of geographic knowledge." Fulfilling this mission, the Society educates and inspires millions every day through its magazines, books, television programs, videos, maps and atlases, research grants, the National Geographic Bee, teacher workshops, and innovative classroom materials. The Society is supported through membership dues, charitable gifts, and income from the sale of its educational products. This support is vital to National Geographic's mission to increase global understanding and promote conservation of our planet through exploration, research, and education.

For more information, please call 1-800-NGS LINE (647-5463) or write to the following address:

National Geographic Society
1145 17th Street N.W., Washington, D.C. 20036-4688 U.S.A.

Visit the Society's Web site at www.nationalgeographic.com.

Library of Congress Cataloging-in-Publication Data

Ballard, Robert D.
Mystery of the ancient seafarers / Robert D. Ballard with Toni Eugene.
p. cm.
Includes bibliographical references and index.
ISBN 0-7922-5845-2 (alk. paper)
Sea Peoples. 2. Mediterranean Region—Antiquities. 3. Mediterranean Region—History—To 476. 4. Middle East—Antiquities. 5. Middle East—History—To 622.
Eugene, Toni. II. Title.

DE73.2.S4B35 2004
930—dc21 2003056163

MYSTERY OF THE ANCIENT SEAFARERS EARLY MARITIME CIVILIZATIONS

Published by the National Geographic Society

John M. Fahey, Jr., *President and Chief Executive Officer*

Gilbert M. Grosvenor, *Chairman of the Board*

Nina D. Hoffman, *Executive Vice President*

Prepared by the Book Division

Kevin Mulroy, *Vice President and Editor-in-Chief*

Charles Kogod, *Illustrations Director*

Marianne R. Koszorus, *Design Director*

Barbara Brownell Grogan, *Executive Editor*

Staff for this Book

Karin Kinney, *Editor and Researcher*

David Griffin, *Art Director*

Sara Elder, *Illustrations Editor*

Carl Mehler, *Director of Maps*

XNR Productions, *Map Research and Production*

ChrisOrr.com, *Map Illustration*

Lewis Bassford, *Production Project Manager*

Sharon Berry, *Illustrations Assistant*

Susan Tyler Hitchcock, *Release Editor*

Mark Wentling, *Indexer*

Elizabeth Booz, *Copy Reader*

Manufacturing and Quality Control

Christopher A. Liedel, *Chief Financial Officer*

Phillip L. Schlosser, *Managing Director*

John T. Dunn, *Technical Director*

Vincent Ryan, *Manager*